PERIPHERAL NERVE REGENERATION - FROM SURGERY TO NEW THERAPEUTIC APPROACHES INCLUDING BIOMATERIALS AND CELL-BASED THERAPIES DEVELOPMENT

Edited by **Ana Colette Maurício**

Peripheral Nerve Regeneration - From Surgery to New Therapeutic Approaches Including Biomaterials and Cell-Based Therapies Development
http://dx.doi.org/10.5772/65612
Edited by Ana Colette Maurício

Contributors

Evgeniya Pushchina, Vicente Vanaclocha, Jose María Ortiz-Criado, Nieves Sáiz-Sapena, Nieves Vanaclocha, Ymer Mekaj, Agon Mekaj, Federica Fregnan, Ofra Ziv-Polat, Michela Morano, Mira M. Mandelbaum-Livnat, Moshe Nissan, Michael Tolmasov, Akiva Korn, Tali Biran, Yifat Bitan, Evgeniy Reider, Nicoletta Viano, Shimon Rochkind, Stefano Geuna, Abraham Shahar, Aysu Hayriye Tezcan, Rafael Cisne, Marcelo Cesar, Rayssa Justo, Marina Manhães, Mauro Geller, Mendel Suchmacher, Pavel Haninec, Libor Mencl

Published by InTech
Janeza Trdine 9, 51000 Rijeka, Croatia

Publishing Process Manager Marijana Francetic
Technical Editor SPi Global
Cover InTech Design team

First published May, 2017
Printed in Croatia
Legal deposit, Croatia: National and University Library in Zagreb

Additional hard copies can be obtained from orders@intechopen.com

Peripheral Nerve Regeneration - From Surgery to New Therapeutic Approaches Including Biomaterials and Cell-Based Therapies Development, Edited by Ana Colette Maurício
p. cm.
Print ISBN 978-953-51-3165-6
Online ISBN 978-953-51-3166-3

For EU product safety concerns:
IN TECH d.o.o., Prolaz Marije Krucifikse Kozulić 3, 51000 Rijeka, Croatia,
info@intechopen.com or visit our website at intechopen.com.

Meet the editor

Ana Colette Maurício has a degree on Veterinary Medicine and a PhD degree on Veterinary Sciences from the Faculty of Veterinary Medicine (FMV)—Technical University of Lisbon (UTL). At the present, she is an associate professor in Habilitation in Veterinary Sciences at Abel Salazar Institute of Biomedical Sciences from the University of Porto (ICBAS-UP), the head of the Veterinary Clinics Department, the director of the Veterinary Sciences Doctoral Program, the vice-president of the Pedagogic Council at ICBAS-UP, and a member of the Ethic Commission for Health Sciences from the University of Porto. She is the scientific coordinator of the Regenerative Medicine and Experimental Surgery research sub-unit from Animal Science Studies Centre (CECA), Institute of Sciences, Technologies and Agro-environment from the University of Porto (ICETA). She is one of the founding shareholders of Biosckin, Molecular and Cell Therapies, SA, a private bank of umbilical cord blood and umbilical cord matrix stem cells authorized by the national authorities, also working in the development of new cell therapies and medical devices for neuro-muscular regeneration (www.criovida.pt). She created a multidisciplinary team allowing a close share of knowledge between biomaterials design, development of cell-based therapies and microsurgery techniques and clinicians' needs. She is the supervisor of several post-graduate students and co-author of a large number of scientific articles and book chapters. She has been a principal researcher of several national and international scientific projects.

Contents

Preface

Peripheral nerve injury still remains a major public health problem with an estimated worldwide incidence of 13–23 cases per 100,000 persons. These injuries may have a traumatic or iatrogenic cause and usually are associated with pain, decrease of function and sensory sensibility with devastating effects on patients' and families' lives. Despite continuous refinement of microsurgery techniques, peripheral nerve repair still stands as one of the most challenging tasks in neurosurgery, as functional recovery is rarely satisfactory in these patients. Poor functional recovery outcome results from nervous system intrinsic and extrinsic factors, such as the integrity of the surrounding tissue post-lesion, the type and level of the injury itself, the effect on the spinal cord and neurons, the compromising of end organs and, with key importance, the timing of the surgery. Also, peripheral nervous system although has spontaneous regeneration ability, there is a very limited prospective of spontaneous recovery, mostly concerning the complete functional neuromuscular recovery.

Direct nerve repair with epineural end-to-end sutures using microsurgery techniques is still the gold standard surgical treatment for severe neurotmesis injuries but only in cases where well-vascularized tension-free coaptation can be achieved. The procedure involves rough fascicular matching between proximal and distal nerve ends and the alignment of nerve fascicles and epineural blood vessels. When peripheral nerve injury originates a significant gap (>3 cm) between the nerve ends with excessive tension for direct epineural repair and reversed interposition, nerve grafts are required. Nerve grafts are single, cable, trunk, interfascicular, or vascularized portions of peripheral nerve with similar diameter to the injured nerve. Nerve grafting may be from autologous or allograph origin. Xenografts have been described as viable alternatives but require extensive immunosuppression, and the transmission of prionic diseases is an important risk, when there is a ruminant donor involved. Nerve autografts are considered the gold standard since they provide appropriate neurotrophic factors and viable Schwann cells, both essential for axonal regeneration without immune compromise. For the choice of the autologous grafts, many factors must be taken into account, such as the size of the nerve gap, location of proposed nerve repair and associated donor-site morbidity.

Tissue engineering was originally defined by Skalak and Fox, in 1988, as "the application of the principles and methods of engineering and life sciences toward the fundamental understanding of structure-function relationships in normal and pathological mammalian tissues and the development of biological substitutes to restore, maintain, or improve functions." Later, in 1993, Langer and Vacanti, in a wide-spread review paper, defined three main pillars of tissue engineering principles: (a) the isolated cells and substitutes—cellular systems, (b) tissue-inducing substances—bioactive molecules, and (c) scaffolds, biomaterials and/or matrices. The first strategy to improve peripheral nerve regeneration concerns cell-based therapies. Stem cells are responsive undifferentiated cells with varying degrees of self-proliferation and differentiation plasticity. Although the number of stem cells is higher before birth, in the adult, there are still several "niches" with a significant number of stem cells. The second pillar focuses on bioactive molecules that can be signalling molecules, proteins and oligonucleotides that can enhance cell migration, cell growth, cell survival and/or differentiation. These bioactive molecules are roughly divided into mitogens, growth factors and morphogens. Finally, the third pillar is the three-dimensional structure that provides shelter and structure for the cellular system. Usually the biomaterials or scaffold mimics the environment and natural extracellular matrix of the place of implantation and should be biocompatible such as their metabolites. Also, scaffolds can be used as drug delivery system in the controlled release of bioactive molecules and pharmaceutical agents. So, recent advances in nerve tissue engineering have greatly promoted the generation of nerve conduits (so-called tube guides) made of several biomaterials, which may be implanted empty or may be filled with growth factors, pharmacological agents and/or cellular systems.

Entubulation offers advantages, including the potential to manipulate the regeneration environment within the tube guide; consequently, guidance of regenerating axons is not only achieved by a mechanical effect but also by chemical, biological and electrical cues. Tube guides provide control environment to outgrowing axons, migration of Schwann cells and neurotrophic stimulation for optimal peripheral nerve regeneration. This approach is usually reserved for gap defects between 1.5 and 3 cm. As early as 1994, Brunelli et al. defined four factors for an ideal nerve conduit material: (i) biocompatibility, (ii) easy preparation and tailoring, (iii) incorporation of neurotrophins and stimulating substances and (iv) protection against scarring. Recently, Arslantunali et al., in 2014, defined the desirable characteristics for a tube guide, as flexibility, biocompatibility, biodegradability, high porosity, neuro-inductivity, neuro-conductivity, easy handling and sufficient mechanical resistance. Nowadays, second- and third-generation nerve conduits are becoming Food and Drug Administration (FDA) approved and reaching the mark, so, many more preclinical and clinical trials are demonstrating major breakthrough in this area.

Multidisciplinary teams, including veterinaries, engineers and medical doctors that through experimental surgery have a crucial role in the development of biomaterials and cell-based therapies, will allow a close share of knowledge between biomaterial design, development of cellular systems and surgeons' needs. For that purpose, the development of new multifunctional tube guides, encompassing biological, chemical, mechanical and electro-mechanical cues, to enhance neuro-muscular regeneration when associated to cellular systems, growth factors and pharmaceutical agents, will greatly improve the functional recovery in peripheral nerve regeneration. Detailed functional and morphological analysis of the regeneration process should also be carried on during the healing period, including a trained surgeons' team, in microsurgery techniques for implantation of the tube guides. Multicompetent research projects in nature will have clinical impact in the improvement of peripheral nerve regeneration, using new biomaterials for the construction of tube guides, associated to innovative therapies, and including a comprehensive and updated functional, morphological and biomechanical analysis of the recovery process.

The present book includes two main sections entitled: (1) peripheral nerve surgery—surgical reconstruction and advances in surgical techniques and (2) biomaterials, neuroprotective factors and cell-based therapies for peripheral nerve regeneration. A comprehensive state of the art of surgical techniques, tissue-engineered nerve graft scaffolds, and their application in nerve regeneration, the advances in peripheral nerve repair and future perspectives will be discussed, including surgeons' and researchers' own large experience in this field of knowledge. Finally, but not less important, I extend my brief and deeply sincere acknowledgement to all the authors that participated in this book and to the supporting team from InTech, especially Ms. Marijana Francetic, Publishing Process Manager.

Ana Colette Maurício, DVM, PhD
Associate Professor in Habilitation
Abel Salazar Institute of Biomedical Sciences
University of Porto (ICBAS-UP)
Porto, Portugal

Section 1

Peripheral Nerve Surgery: Surgical Reconstruction and Advances of Surgical Techniques

Chapter 1

Peripheral Nerve Injury and Current Treatment Strategies

Aysu Hayriye Tezcan

Additional information is available at the end of the chapter

http://dx.doi.org/10.5772/intechopen.68345

Abstract

Neuronal cells are the main fundamental anatomic unit of the system. Nerve injuries are generally divided into three categories as neuropraxia, axonotmesis and neurotmesis. Neurotmesis is the most severe form. Schwann cells are activated within 24 hours of the injury and the healing cascade continued with the cells, which are stimulated by Schwann cells. And neurotrophic factors like nerve growth factor (NGF) have a crucial role in regeneration and degeneration processes. Additionally, Schwann cells upregulate the expression of some proteins, such as fibronectin, which are crucial for axonal regeneration. All this information about nerve healing sheds light on treatment studies. Iatrogenic nerve injury has an important place in peripheral nerve injury. Causes may be direct surgical damage, wrong intraoperative patient positioning, anaesthesia-related reasons or limb tourniquets. Typical symptoms are motor or sensory deficits such as paraesthesia, weakness, paralysis and pain. Many of the traumatic nerve injuries require surgical repair. Direct nerve repair and autologous nerve grafts are still gold-standard treatment options. Additionally, nerve conduits are very successful to provide an ideal peripheral support for neuronal recovery but are still insufficient. In recent years, research efforts have focused on the neurotrophic factors and cell-based therapies to perform better microenvironment for neuronal healing.

Keywords: peripheral nerve, injury, iatrogenic, nerve grafts, conduit, cell-based therapy

1. Introduction

An injury to a nerve can result in a problem with the muscle innervation or in a loss of sensation. In some people, it can also cause pain. The type of nerve injury will determine the type of treatment that will be needed. Peripheral nerve injuries (PNIs) affect all age groups and have many causes like trauma and medical disorders. The majority of the peripheral nerve injuries

(PNI) occur in the upper extremity and are secondary to trauma [1]. Typical symptoms are motor or sensory deficits such as paraesthesia, weakness, paralysis and pain [2]. Many of the traumatic nerve injuries require surgical repair. The primary goal of the repair is to achieve the reinnervation of the target organs, but the therapeutic options which are used at present cannot achieve perfect sensory and motor recovery in all cases. In this chapter, the pathophysiology and mechanisms of nerve injuries, the traumatic nerve injuries, especially the iatrogenic forms, and therapeutic options for the peripheral nerve injuries will be discussed.

2. Peripheral nerve anatomy

Peripheral nervous system consists of neuronal cells, glial cells and stromal cells. Neuronal cells are the main fundamental anatomic unit of the system. Neurons conduct electrical signals between central nervous system and other parts of the body. Afferent neurons carry information from the body to the brain, and efferent neurons carry signals from central nervous system to target organs. Each neuron consists of a cell body, dendrites and an axon. Transportation of the action potential is from cell body through the extension of the body called axons. Cell body is the source of nutritional elements and neurotransmitters. Nerve fibres are categorized into three groups based on their functions (motor, sensory and autonomic neurons) and into two groups according to structural characteristics (myelinated and unmyelinated). Myelin is a critical molecule for conduction of action potential, which is composed of 30% proteins and 70% lipids. Schwann cells (SCs) constitute the major neuroglial component of the peripheral nervous system and surround axons at regular intervals forming the myelin. And there are unmyelinated intersegmental areas known as Ranvier nodes. Ranvier node is enriched with voltage-sensitive sodium channels and generates ionic impulses only at the node. Owing to this structure, action potential jumps from one node to the next along the length of the axon and saltatory conduction occurs. This process results in faster conduction of the action potential in myelinated nerves rather than unmyelinated nerves [3, 4].

Connective support of the neuron consists of three laminary structures including endoneurium, perineurium, epineurium and mesoneurium. The deepest structural layer nearest to the axon is endoneurium, which is composed of collagen fibres. Perineurium surrounds multiple nerve fibres and their endoneuriums. Perineurium is composed of collagen fibres and perineural cells, which provide a protective barrier between the nerve and its blood supply. The entry of large proteins, toxins, antigens and infectious agents is prevented by that barrier (blood-nerve barrier) [3]. Perineurium has a major role in maintaining the integrity and providing tensile strength and elasticity of the nerve [5]. The external layer is the epineurium, which consists of two components. Interfascicular epineurium surrounds each nerve fascicle, and the extrafascicular epineurium invests the entire nerve trunk and clinches the blood vessels to the nerve [4]. The epineurium is a quite strong layer and resists to compression injuries with its collagen fibres and lipid globules. Besides, epineurium is known as the most abundant type of connective tissue of the nerve. Eighty-eight per cent of the sciatic nerve is composed of epineurium at the gluteal level [6]. The outermost tissue around the nerve trunk is mesoneurium. Mesoneurium harbours anastomoses between veins and arterioles of the nerve. Mesoneurium limits the frictional forces during movement of the nerve against closed local structures [7].

3. Grading nerve injury

The first approach to the nerve injury is grading because that defines the severity of the injury and success of the repair. The most widely accepted grading system, which was defined by Seddon and Sunderland, is according to the microscopic changes secondary to the injury [6, 8].

Seddon divided nerve injuries into three categories as neuropraxia, axonotmesis and neurotmesis. The mildest form of nerve injury is neuropraxia, which is characterized by focal demyelination without damage to nerve continuity. This kind of injury does not cause distal degeneration. Neuropraxia typically occurs after compression or traction of the nerve. The conduction velocity is decreased, but the damage is transient [9]. The complete recovery time of the injury varies from 1 week to 6 months [10]. Axonotmesis is the damage to the axons with focal demyelination where connective tissues (perineurium and epineurium) of the nerve are preserved. Distal axon and myelin degeneration causes complete denervation. Although the motor and sensory function recovery is not as fast as in neuropraxia, preservation of the supportive tissue makes recovery perfect in axonotmesis. The most severe form of the nerve injury is neurotmesis, which is defined as full anatomical and physiological transection of the nerve. All functional components are damaged, and recovery without surgical repair is impossible [11].

Sunderland divided nerve injuries into five categories mostly according to supportive tissue damage [6]. First-degree injury is equivalent to neuropraxia defined as partial disruption in conduction. Second-degree injury is equivalent to axonotmesis. In the third-degree injury, perineurium is left intact, but endoneurium and axon are disrupted. The functional recovery of the nerve depends on the extent of the injury. In the fourth-degree injury, epineurium is the only structure that remains intact. Surgical repair is required for recovery. The fifth-degree injury defines complete transaction of the nerve, and as in neurotmesis, surgical repair is necessary [4].

4. Pathophysiological changes after nerve injury

The injury to the neuronal tissue does not result in mitosis or cell proliferation like other tissues of the human body. First-degree injuries result in mild or no pathological changes. Only decrement of conduction velocity is observed, and none of the regeneration or degeneration processes are activated. However, in the second-degree injury, a unique pathophysiological change defined as Wallerian degeneration occurs in the neural tissue which is not observed in any other tissues [12]. In Wallerian degeneration, a calcium-mediated process results in anterograde degeneration distal to the injury site. Wallerian degeneration ensues 24–48 hours after the injury with fragmentation of distal axons and myelin is formed histologically [13]. Disorganization of neurotubules and neurofilaments causes irregular axonal margins. Structural axonal durability is lost 48–96 hours after injury and conduction of impulses ceases. Myelin disintegration lags slightly behind that of axons but is well advanced by 36–48 hours [11].

Schwann cells are activated within 24 hours of injury. Their initial role is phagocytosis of the axonal and myelin debris. Macrophages migrate to the injury site and stimulate proliferation of Schwann cells and fibroblasts. Schwann cells and macrophages work together to clean up the injury site which may take 1 week to several months. The other significant role of Schwann cells is to complete the cleaned endoneurial tubes in organized longitudinal columns defined as Bungner's bands. Bungner's bands are important guides for sprouting axons during reinnervation [11, 14, 15].

The third-degree injury (intrafascicular injury) involves retraction of the severed nerve fibre ends due to elastic endoneurium. Local trauma leads to a significant inflammatory response. Fibroblast proliferation aggravates the process, and a dense interfascicular scar forms. This kind of injury impairs axonal regeneration, and endoneurial tubes remain denervated. If the endoneurial tube does not receive a regenerating axon, progressive fibrosis eventually obliterates it. In the fourth- and fifth-degree injuries, activated Schwann cells and fibroblasts cause vigorous cellular proliferation. Local vascular trauma leads to macrophage accumulation. In these injuries, the nerve ends become an irregular mass of Schwann cells, fibroblasts, macrophages and collagen fibres. Regenerating axons reach that disorganized proximal stump and encounter a rough barrier that impedes further growth [11].

Disconnection of cell bodies and axons activates programmed cell death pathway within 6 hours of the injury in a process called chromatolysis [9, 15].

Neurotrophic factors have a crucial role in regeneration and degeneration processes. Macrophages that migrate to the injury site express interleukin-1 (IL-1). IL-1 stimulates nerve growth factor (NGF) production by Schwann cells. NGF is very important for axonal regeneration process and myelin formation [16]. Additionally, Schwann cells upregulate the expression of fibronectin, laminin and neurotrophins, which are crucial for axonal regeneration [11]. All of these factors are not enough for recovery of transected nerves. If the surgical repair fails, denervation of the distal end leads to lack of impulse conduction followed by loss of motor end plates and muscle fibrosis [17].

5. Mechanisms of peripheral nerve injury

There are various mechanisms that lead to peripheral nerve injury. All types of traumatic injury can have an iatrogenic component. Direct nerve damage is one of the mechanisms that may occur during surgery or accidentally secondary to an external trauma. These iatrogenic injuries may be provoked by knives, propellers and scalpels used in surgery or a needle trauma secondary to the anaesthesia technique. Traction injuries may occur during surgery by surgical retractors or wrong positioning of the limbs. Compression injury occurs secondary to wrong positioning of the limbs, poor padding of the body surfaces or use of surgical tourniquets. The pathophysiological mechanisms of compression injury involve both mechanical compression damage and ischemic damage. Ischemic injury may be caused by surgical tourniquets, prolonged immobility, haematoma surrounding a nerve or vasoconstrictor agents. Injection of neurotoxic drugs is another iatrogenic cause of nerve injury. An additional mechanism involved in peripheral nerve injury is double crush syndrome. Patients who have medical comorbidities (e.g., diabetes mellitus, rheumatoid arthritis) associated with peripheral neuropathy have more sensitive nerves and are more susceptible to a secondary damage like compression or laceration. A minor injury may cause permanent nerve injury in these patients. Patients with features such as diabetes mellitus, obesity, peripheral vascular disease, arthritis, alcohol usage, tobacco usage, advanced age, circulatory failure, extreme weakness are more likely to have nerve damage in the perioperative period [18–30].

There are also other risk factors which are independent from patients. Neurosurgery, cardiac surgery, gastrointestinal surgery and orthopaedic surgery are associated with a higher incidence of peripheral nerve injury. Perioperative risk factors include hypovolemia, hypotension, hypoxia, electrolyte disturbances, hypothermia, length of surgery (>2–4 hours) and patient positioning [19, 21, 31–35]. General anaesthesia and neuraxial blockage have a risk for peripheral nerve injury because they limit patients' own position changes as compared to mildly sedated patients.

6. Causes of iatrogenic nerve injury

In one study, 17.4% of the traumatic nerve injury cases ($n = 722$) surgically treated in a tertiary care service were iatrogenic in origin. And 94% of them occurred during a previous operation [36]. In an extensive retrospective study spanning a period of 10 years, the frequency of perioperative peripheral nerve injuries was 0.03% [37]. When operated nerves were examined individually based on surgical specialty, 25% of the operated sciatic nerve injuries, 50% of the operated femoral nerve injuries and 93% of the operated accessory nerve injuries were found to be caused by iatrogenic injuries [38–40].

Iatrogenic nerve injury may be associated with operational or nonoperational factors. Needle injections and external compressions are common nonoperational causes. Operational factors may be directly related to surgery or may involve wrong patient positioning. Direct intraoperative injury mechanisms include cutting, crushing, tying off, penetrating, twisting by screws, stretching by retractors, grabbing or squeezing by repositioned bone, coagulating with cautery or bipolar forceps, burning by cement or excision of the target pathology. It was reported that 94% of the iatrogenic peripheral nerve injury operations were caused by direct surgical damage [36, 41–43].

Nerves cannot always be clearly visualized in the operative field. There are reports that they have been mistakenly recognized as a tendon or vessel [44–47]. And sometimes nerves may be removed during nerve sheath tumour or lymph node excision (accessory nerve may be damaged during lymph node dissection from posterior triangle of the neck).

Nerve sheath tumours are another surgical risk factor for nerve injuries because they are rarely truly recognized and approached. They may be diagnosed as an unspecified lump, undetected tumour or ganglion, so the surgical interventions may result in disappointment.

Nerve sheath tumours are mostly of benign nature (schwannoma or neurofibroma), and appropriate surgical interventions are able to correct neurological deficits [48].

In order to avoid such outcomes, surgeons must have extensive anatomic knowledge about peripheral nerves and be experienced in specialized surgical intervention.

6.1. High-risk surgical procedures

In addition to all other external factors, certain surgeries carry a high risk for peripheral nerve injury. Common procedures that may cause nerve injury are osteosynthesis, osteotomy, arthrodesis, lymph node biopsies from the posterior triangle of the neck, carpal tunnel release, Baker's cyst excision, varicose vein surgery and inguinal hernia repair [41].

A study reported that the most common surgical interventions associated with peripheral nerve injury are as follows in decreasing order: 45% major procedures (orthopaedic procedures, abdominal surgery and trauma), 27% minor procedures (lymph node surgery and varicose vein surgery), 15% neurosurgical interventions (carpal tunnel operations and tenolysis), and 4% non-surgical interventions (arterial or venous puncture and plaster cast) [36, 41].

6.2. Anaesthesia and peripheral nerve injury

Anaesthesia has a role in the emergence of perioperative peripheral nerve injury. Both general and regional anaesthesia may potentially cause peripheral nerve injury with different mechanisms including compression, stretch, direct nerve trauma and local chemical toxicity [49]. An extensive retrospective review reported the incidence of perioperative PNI at 0.03%. The same review demonstrated that general anaesthesia and epidural anaesthesia have associations with nerve injury, but peripheral and spinal nerve blocks do not [37]. The ulnar nerve is the most commonly compromised nerve under anaesthesia [50]. Predisposing factors that make patients more susceptible to nerve injury are diabetes mellitus, hypertension, tobacco use, arthritis, obesity and low body weight [28, 37]. Except for compression injuries secondary to improper patient positioning, direct needle trauma may be the main cause of the anaesthesia-related peripheral nerve injury. There are various risk factors that affect the risk of nerve trauma with regional anaesthesia. Needle may damage the nerve fascicles directly, and damaged nerve vessels and existing extraneural or intraneural haematoma may compromise nerve fascicles [51]. Injecting the local anaesthetic into a fascicle is the main source of peripheral nerve block-related nerve injury. Intrafascicular injection traumatizes the nerve, damages perineurium and results in the loss of protective environment within the fascicle [37, 51–53]. Epinephrine when combined with local anaesthetics tends to cause local vasoconstriction, but its role in causing nerve injury is still controversial [54]. High-pressure local anaesthetic injection may damage neuronal microvasculature and cause neural ischemia [55]. Both neuraxial block and peripheral nerve blocks' needle type is associated with neuronal injury risk. Nerve puncture with pencil-point or Tuohy needles causes a similarly high degree of injury [56]. Human studies could not demonstrate the difference between neurostimulation-guided or ultrasound-guided nerve blocks in terms of nerve injury [51].

Fortunately, complications of neuraxial blocks are rare. A large-scale research study reported that the incidence of neuraxial complications is 0.0075%, and 67% of them results in permanent injury. The most common complications are spinal hematoma, cauda equina syndrome, purulent meningitis and epidural abscess. Studies mentioned that the risk of spinal haematoma is higher with epidural anaesthesia compared to spinal anaesthesia [57]. The risk is higher in patients who have coagulation abnormalities, advanced age, female gender, concurrent spinal stenosis or preexisting neurological diseases [58].

Improper positioning is another cause of anaesthesia-related perioperative peripheral nerve injury. Wrong patient positioning increases the compression pressure on nerves which are superficial or in close proximity to a bone. For example, the ulnar nerve is a superficial nerve which is in the vicinity of the medial epicondyle of the humerus, or the peroneal nerve is near the fibular neck. Additionally, excessive tractions of the nerves may be another injury mechanism (excessive traction of the shoulder causes brachial plexus injury). Anaesthesiologists should be aware of the body position and its association with nerve injury (e.g., lithotomy and femoral nerve injury, lateral decubitus position and peroneal nerve injury, taping the shoulders and brachial plexus injury) [41]. The American Society of Anaesthesiologists published a practice guideline for prevention of positioning-related peripheral nerve injury. The guideline recommends careful positioning of the patient, protective padding, padded arm boards and avoidance of contact with hard surfaces. Specifically, the guideline advised that arm abduction should be limited to $<90°$ in the supine position to protect brachial plexus, neutral forearm positioning or supination of the hand should be done with padding in the supine position to protect ulnar and median nerves, flexion of the elbow should be limited to $<90°$, appropriate padding should be applied in lithotomy in lateral and prone positioning, and hip flexion should be $<120°$ in lithotomy positioning [59].

6.3. Tourniquet-related nerve injury (TRNI)

Despite its rare but devastating side effects, pneumatic surgical tourniquets are widely used in orthopaedic and plastic surgery because bloodless operative site is very important for optimum surgery. Pathophysiological mechanisms of the TRNI include mechanical compression and neural ischemia [60]. Mechanical compression causes microvascular congestion, inadequate perfusion, axonal degeneration and transient loss of innervation function [61]. These effects usually start within 2–3 hours. Even when the insufflation pressure is appropriate or tourniquet time is short, the potential for nerve injury still persists [62]. Fortunately, when standard recommendations are followed, major neurological deficit is rare. The incidence of permanent injury is 0.032% [63]. If the mechanical stress is the major pathological factor in TRNI, it is recommended that the placement of the tourniquet be as proximal as possible to the limb to enlarge the limb circumference with muscle mass with the aim to protect the nerves [64].

6.4. Regions with higher risk of nerve injury

In some body regions, nerves lay superficial and have close proximity with hard tissue. In these superficial regions, nerves have a narrow diameter that makes them more susceptible to the injury. These sites include the wrist (ulnar nerve), posterior triangle of the neck (spinal accessory nerve (SAN)), posterior of the knee and popliteal fossa, fibular head (peroneal nerve), elbow and groin (ilioinguinal, iliohypogastric nerve) [8].

7. Peripheral nerves and clinical presentations of injuries

Based on the results of a large clinical experience report, the median nerve is the most commonly injured nerve among the iatrogenic nerve injuries. It is followed by the accessory nerve, radial nerve, common peroneal nerve (CPN), ulnar nerve and femoral nerve [41]. The diagnosis of iatrogenic nerve injury is easy. An asymptomatic patient starts to complain about neurological symptoms after a surgical intervention. Neurological symptoms may include sensory deficits or motor movement limitations. Sensory symptoms may include anaesthesia, paraesthesia, hypoesthesia, hyperesthesia and pain. Motor limitations may include paresis and paralysis. Autonomic dysfunction or neuropathic pain may be observed depending on the type of nerve and the injury process [49].

7.1. Lingual and inferior alveolar nerves

The lingual nerve is a sensory nerve, the inferior alveolar nerve is both sensory and motor, and two of them are branches of the mandibular nerve. These nerves are mostly injured during orodental practices such as exodontic and endodontic procedures, dental implants and injection, and osteotomy. Injury symptoms are mostly similar to those mentioned above. Paraesthesia is a common symptom experienced by patients with these nerves injuries. The common cause of iatrogenic paraesthesia is extraction of third molars and dental injections. Twenty-five per cent of patients with iatrogenic paraesthesia suffer from permanent conditions and 75% regain normal sensation spontaneously without further treatment within 6–8 weeks [65, 66]. Spontaneous recovery may be prolonged up to 24 months. Microsurgery may be indicated in the following cases: confirmed transection of a nerve, total anaesthesia of the affected area 2 months after the trauma, lack of protective reflexes (on biting or burning of the tongue or lower lip) 2 months after trauma, dysaesthesia. Considerable functional improvement after surgery may be seen, but regaining normal sensation is not always possible [67].

7.2. Laryngeal nerves

The internal branch of the superior laryngeal nerve (IBSLN) provides general sensation for the tissue superior to the vocal cords. The external branch of the superior laryngeal nerve (EBSLN) innervates cricothyroid muscle and contributes to innervation of the pharyngeal plexus. The ESBLN is closely related to the superior thyroid vascular pedicle, and this anatomic relationship makes the nerve vulnerable to iatrogenic injuries. The recurrent laryngeal nerve (RLN) supplies four intrinsic muscles of the larynx except cricothyroid muscles and supplies the mucosa of vocal cords and subglottis. Although controversy still exists, exploration and dissection of the nerve are still recommended to avoid nerve injury during surgery. Thus, surgeons should identify some landmarks (inferior thyroid artery, tracheoesophageal nerve, Berry's ligament, Zuckerkandl's tubercle) which are closely related to the RLN to help protect the nerve from injury [68].

Iatrogenic laryngeal nerve injury is mostly seen after anterior neck surgery (thyroid, parathyroid, anterior cervical spine surgeries, carotid endarterectomy). While advanced neurophysiological monitoring techniques are performed during anterior neck surgery, clear visualization of the nerve during surgery is still more important to prevent nerve injury [69]. Symptoms of laryngeal nerve injury are difficulty in speaking, difficulty in swallowing, hoarseness and breathing problems (if the injury is bilateral) [68]. Laryngeal electromyography (EMG) has been the gold standard in the diagnosis of laryngeal nerve injuries. Iatrogenic RLN injury was found in 0.3–13.2% of the cases, and superior laryngeal nerve (SLN) injury occurred in less than 5% of the cases. Transient vocal cord palsy recovers within 6 months. Permanent vocal cord palsy, which is not recovering within 1 year, occurs in 0.4 and 2.8% of the cases [70–76].

The treatment of RLN injury includes medicines (neurotrophic agents, glucocorticoids or vasodilators), ultrashort wave therapy, voice training, vocal cord injections and reinnervation methods (decompression, end-to-end anastomosis, implantation of an ansa cervicalis nerve-muscle pedicle). Recent treatment methods include thyroplasty, laser arytenoidectomy and cordectomy. The recovery success of the RLN depends on early diagnosis and early exploration of the nerve [77–79].

The treatment of SLN injury includes voice therapy, type 1 or 4 thyroplasty and reinnervation using nerve muscle pedicle technique. There are no studies which extensively investigated the success of these therapeutic methods for SLN injury [80–83].

7.3. Facial nerve

Published facial nerve injury rates in oral surgical procedures vary from 2 to 25% and 0.6 to 3.7% in primary tympanomastoidectomy. Secondary tympanomastoidectomy procedures doubled the risk of iatrogenic facial nerve injury (IFNI) (4–10%). In an epidemiologic report, oral maxillofacial surgery was the most common procedure that caused iatrogenic facial nerve injury. Temporomandibular joint reconstruction, mastoidectomy, parotidectomy and rhytidectomy are the most risky procedures in terms of IFNI. Hemifacial paralysis is the most prevalent pattern of IFNI. When IFNI is diagnosed in a patient, he/she must be referred for facial nerve exploration immediately. Reconstruction should be performed within 4–6 months after surgery to avoid severe atrophy of the mimetic muscles that occurs 12 months after denervation. Currently, no medical treatment exists for facial nerve injury. Systemic corticosteroids have minimal contribution to the recovery, and the mainstay of treatment is surgery. Surgical options include direct repair, cable nerve grafting and nerve substitution techniques. Residual weakness and synkinesis are common results of facial nerve repair. Common factors restricting functional recovery include older age, long grafts and extended delay between injury and repair [84–89].

7.4. Spinal accessory nerve

Cervical lymph node biopsy is the most common cause of iatrogenic spinal accessory nerve (SAN) injury, and the incidence after this procedure varies between 3 and 10%. SAN injury results in the loss of trapezius muscle motor function and weakness of the shoulder abduction, dropping of the shoulder, winging scapula and shoulder pain. Although SAN is a motor nerve, patients often describe a sharp, electric shock pain during the procedure. Spontaneous recovery of SAN injury is very rare. If the lesion is left untreated, pain or functional deficit will occur in 60–90% of patients. The surgical repair procedures include end-to-end anastomosis and nerve grafting. Early reconstruction surgery (within 3–6 months of injury) is recommended to avoid permanent functional deficit of the trapezius muscle. However, positive functional results can still be expected within 9 months of the injury since the distance to the target muscle is short [40, 89–92].

7.5. Brachial plexus

The anatomical features of the brachial plexus make it susceptible to iatrogenic injuries. Brachial plexus is superficial, lies between mobile bones and has a limited range of movement between the clavicle and the first rib. Iatrogenic injuries may be induced by compression, stretching or direct damage. Median sternotomy during cardiac surgery, lateral decubitus positioning, shoulder surgery, arm abduction and shoulder displacement are the iatrogenic causes of the injury. Motor dysfunction is the major clinical presentation that depends on the injury level. Surgical treatment options include neurolysis, nerve suture, nerve grafting and neurotization. Timing of surgery differs according to the injury type. Laceration injuries should be explored acutely, blunt injury reconstruction may be performed within 2–3 weeks of the injury and closed traction injury is operated within 4–5 months of the surgery [49].

7.6. Median nerve

As demonstrated in a comprehensive research study, median nerve is the nerve most commonly susceptible to iatrogenic nerve injury [42]. Median nerve is mostly harmed during carpal tunnel release procedures. Surgeon's knowledge on skin landmarks may diminish the nerve damage. Median nerve injury presents with paraesthesia in the palmar side of the fingers, weakness of abduction, opposition of the thumb and forearm being kept in supination. Near the motor deficit, patient may complain about electric-like shooting pain. Early exploration and treatment are recommended before chronic pain syndrome is manifested [49].

7.7. Ulnar nerve

Ulnar nerve injury is the most common perioperative peripheral nerve injury because the ulnar nerve is superficial and in close proximity to the medial epicondyle of the humerus. Subclinic neuropathy is the most common presentation. Tingling, numbness along the little finger, weakness of abduction and abduction of the fingers are usual clinical presentations [49].

7.8. Radial nerve

Radial nerve lies along the spiral groove of the humerus, and this location is the most injured site of the nerve by dislocated humeral fracture. Open reduction and fixation of the fracture may cause iatrogenic nerve injury. Kirschner wire placement for radial fractures can penetrate the nerve. Thus, mini-incisions for surgical procedures rather than percutaneous procedures are preferable to protect the nerve. There are numerous treatment options for radial nerve repair. None of them has proven better outcomes than others. However, radial nerve has a perfect recovery profile after reconstruction [49].

7.9. Phrenic nerve

The phrenic nerve supplies diaphragm that originates from cervical nerve roots and descends through the thorax. Left phrenic nerve lies within the pericardium. The phrenic nerve paralysis occurs mostly after cardiothoracic surgery with an incidence of 1.4–7%. In addition to trauma, pericardial crushed ice placement is another cause of iatrogenic nerve injury in cardiac surgery. Elevated hemidiaphragm, lower lobe atelectasis and poor respiratory effort are the clinical presentations of the paralysis. Cold injury and internal mammary artery dissection increase the risk of phrenic nerve injury. Even if it takes weeks and months, the prognosis of phrenic nerve injury is good. Most patients recover within 1 year eventually [93–95].

7.10. Inguinal nerves

The ilioinguinal, genitofemoral and iliohypogastric nerves are cut, coagulated, sutured or incorporated into a mesh during open and endoscopic inguinal hernia repairs. These nerves may also be injured during laparotomy. Inguinal nerve injury during hernia repair occurs in 0.5–2% of the procedures. Clinical presentation of these nerve injuries includes a sharp, burning pain radiating to the suprapubic area, labia or scrotum, paraesthesia over the same areas and pain relief after infiltration of a local anaesthetic. Symptoms are aggravated by stretching, coughing, sneezing and Valsalva manoeuvres. Pain is diminished in more than 90% of the patients after neurectomy and excision of the injured nerve [96–100].

7.11. Femoral nerve

Postoperative hematoma, cement extrusion, trauma from retractors, bone or prosthesis malpositioning and lithotomy position with extreme abduction of thighs are the main causes of iatrogenic femoral nerve injury. Deep pelvic surgery and abdominal surgeries are procedures that are most commonly implicated in iatrogenic femoral nerve injury. In these operations, the common risk factor is compression by retractor blades. Thin subcutaneous fat layer, surgical duration longer than four hours, poorly developed rectus muscle, narrow pelvis and self-retaining retractors are the risk factors for femoral nerve compression. Clinical presentations of femoral nerve injury include loss of sensation at the front of the thigh, weak hip flexion and loss of knee extension. Although severe nerve injuries need long nerve grafts, femoral nerve reconstructions result in good functional recovery [101–103].

7.12. Sciatic nerve

Sciatic nerve injury manifests itself as paralysis of the hamstring muscle, foot drop and impaired sensation below the knee except medial regions. The lithotomy, frog leg and sitting positions cause perioperative nerve injury by stretching the nerve [49]. Sciatic nerve injuries are mostly reported as a complication of total hip arthroplasty with an incidence of 0.16–8%. Hip dysplasia, posterior approach, revision surgery (3.2%), limb lengthening, female gender and younger age are the primary risk factors of nerve injury. The most common mechanisms include tension, direct trauma by retractors, compression by postoperative hematoma, direct lacerations and sutures. Sciatic nerve injury occurs in 1.7% of hip arthroscopies. Complete injuries without therapy cause 100% limb disability [104–108]. Results of operative treatments are mixed. A large series of sciatic nerve repairs reported that tibial division recovery is better than peroneal division, and outcomes at the thigh are better than at the buttock [101].

7.13. Common peroneal nerve

The common peroneal nerve (CPN) injury presents with loss of dorsiflexion and eversion of the foot (equinovarus deformity). Sensory impairment occurs in the anterolateral region of the leg. Iatrogenic injuries are related to high tibial osteotomies (HTO) (4.9%) and total knee arthroplasties (TKA) (0.3–9.5%). Revision surgeries, preoperative valgus deformity, rheumatoid arthritis, history of previous HTO, prolonged tourniquet time and history of laminectomy are the risk factors of perioperative CPN injury [109–113]. Lithotomy and lateral position are the nonsurgical perioperative risk factors for CPN injury which is associated with fibular head compression. Full recovery rates of partial nerve injury vary between 76 and 87%, and complete injury recovery is seen in 20–35% of the cases [101]. Surgical treatment is indicated if no evidence for recovery is observed.

8. Functional assessment and management of peripheral nerve injury

Clinical presentations of peripheral nerve injury usually include pain, dysaesthesia and partial or complete loss of motor and sensory functions. Evaluation of the injury starts with complete history and physical examination. Diagnosis is established through electrodiagnostic assessment and radiological studies. Treatment options are performed according to the wound characteristics at the appropriate timing in terms of the injury mechanism after all examinations.

Physical examination should involve detailed assessment of motor and sensory deficits. Moving and static two-point discrimination, sharp and dull discrimination, grading of grip and pinch strength should be tested and recorded preoperatively. Subsequent to these procedures, peripheral tissues of the injury site (soft tissue, vascularization) should be examined. Tinel's sign over the course of the injured nerve will be positive. It refers to paraesthesia elicited by lightly tapping over the suspected location. Tinel's sign is elicited by regenerating axonal growth [114].

The timing of functional assessment according to the injury type is an important issue for diagnostic success. Optimum timing for electrodiagnostic studies varies between injury types. The formation of Wallerian degeneration is the main point of distinction of the neuropraxia, axonotmesis or neurotmesis. Initial electrodiagnostic studies 7–10 days after an acute injury may be helpful to localize the lesion and to distinguish conduction blocks from axonotmesis. Electrodiagnostic studies, which are performed 3–4 weeks after surgery, provide more information about the lesion because fibrillation potentials may not appear until this time [115]. Electrodiagnostic studies are mostly useful during intraoperative management while checking an action potential distal to the injury. They are also useful to distinguish the intact fascicle and neuroma intraoperatively. Electrophysiological procedures have a critical role in monitoring recovery and determining neuronal reinnervation 2–4 months after nerve surgery [114].

Electromyography (EMG) demonstrates the electrical activity of a muscle and is useful for detecting injuries of efferent peripheral neuron and motor unit. Reduced number of functional axons after an injury causes EMG abnormalities. In neuropraxia, it demonstrates normal or decreased recruitment. However, following lesions causing axon loss, EMG demonstrates abnormal activity such as fibrillation and positive sharp waves. EMG may be normal immediately after axonotmesis and demonstrates abnormal activity after 10–14 days. The onset of the fibrillations depends on the length of the distal nerve stump. Fibrillation development with short stumps may take 10–14 days, and for longer stumps, it may take 21–30 days (e.g., ulnar-innervated hand muscles in a brachial plexopathy) [115].

Nerve conduction studies (NCS) assess both motor and sensory functions of the nerve via a voltage stimulator. The evoked response is recorded from a surface electrode overlying the muscle (motor response) or nerve (sensory response). NCS is used initially to demonstrate the presence of conduction block [13]. In pure neuropraxic injuries, motor response failures occur immediately and conduction is detected to be normal distal to the lesion. Electrodiagnostically complete axonotmesis and complete neurotmesis demonstrate the same results. Immediately after axonotmesis motor conduction studies show the same pattern as neuropraxia until Wallerian degeneration has occurred. Typically, Wallerian degeneration occurs 9 days after injury, and it becomes possible to distinguish neuropraxia and axonotmesis. After this time, amplitude of the motor response falls and later responses are absent in both proximal and distal of the lesion [115]. Consequently, EMG and NCS should be used complementarily to define the characteristics of the injury (complete or incomplete injury, localization, injury age, injury grade, prognosis and postoperative recovery) [49].

Radiological studies have a limited role in diagnosing peripheral nerve injuries. However, they may be used complementary to the electrodiagnostic studies. Mostly magnetic resonance imaging (MRI) and high-resolution ultrasound are used for peripheral neuroimaging. Both are helpful to determine the exact site of the lesion particularly when electrodiagnostic studies are insufficient [49].

9. Treatment strategies of peripheral nerve injury

The recovery time of the injured nerve depends on various external factors including most importantly early nerve exploration and repair. However, it should be known that axonal regeneration rate is as slow as 1–2 mm per day and there is no treatment to accelerate this process [15]. The irreversible motor unit degeneration starts 12–18 months after denervation of the muscle but may persist for 26 months [116]. Recovery of sensory regeneration may take longer. Additional injury at target muscles or an injury in peripheral supportive tissue delays the recovery more than usual.

The features of the injury define the type or timing of the surgical nerve repair. There are three types of wound including tidy, untidy and closed traction injuries. The tidy wound may be made by a glass or a scalpel, which has sharp edges and primary repair is a preferable treatment option. Untidy wound samples are open fractures or gunshot wounds with extensive tissue damage and infection and cannot be repaired immediately. Closed traction injuries have retracted and damaged nerves, vessels and peripheral supportive tissues. Closed traction injuries have the worst outcomes of all wounds.

Nerve exploration and repair indications are paralysis around the injured nerve, closed injury with supportive tissue damage, open injury requiring open reduction and internal fixation, nerve lesions with arterial damage, traction injuries to the brachial plexus, declining nerve function after diagnosis, failure of neurological improvement, failure to improve after a conduction block within 6 weeks of the injury and persistent pain or neuroma formation [4].

9.1. Direct nerve repair

Direct nerve repair with microsurgical techniques is still a gold-standard treatment method for axonotmesis and neurotmesis injuries [14]. Direct suturing repair without grafting is used for short nerve deficits (<5 mm). Larger deficits require nerve grafting. Repair of a large gap without grafting exhibits excessive tension and produce poor outcomes. In terms of recovery for larger gaps, repair with nerve grafting has better outcomes compared to primary repair without grafting [4].

9.2. Fibrin glue

Fibrin glue enables primary sutureless repair with an adhesive. Repair with fibrin sealants ensures shorter recovery time, less fibrosis and decreased inflammatory reactions. Studies do not report any difference between fibrin glue and direct suturing in terms of axonal regeneration, fibre alignment and nerve conduction velocity recovery [117]. The most important advantage of the fibrin glue is quick and easy application in emergency conditions whenever there is absence of experienced surgeon for nerve repair.

9.3. Nerve grafts

Autologous nerve grafts are the gold-standard option in peripheral nerve repairs. As reported in the literature, autologous nerve grafting has better recovery results in long nerve deficits (>3 cm), more proximal injuries and critical nerve injuries [15]. Donor nerve grafts are extracted from expandable sensory nerves such as sural and medial antebrachial nerves. Autologous nerve grafting has best results because it has necessary materials for nerve regeneration like Schwann cells, basal lamina, neurotrophic factors and adhesion molecules [2]. Despite its superior results, using autologous nerve grafts has some limitations including limited tissue availability, necessity for second surgery, the graft, donor-site morbidity, loss of nerve function and potential difference in tissue size. Allograft nerves that are collected from a cadaver or donor for nerve grafting are other options. Donor allografts contain viable donor Schwann cells, and immunosuppressive therapy is needed for 18–24 months to inactivate these cells for sustained regeneration. Immunosuppressive therapy has many side effects including opportunistic infections and tumour formation. Cadaveric nerve allografts are used in patients who have inadequate autologous nerve grafts. The recovery results are good, but the process is too expensive and requires experience to perform [118–121].

Currently, scientists focused their research on acellular human nerve allografts with the aim to eliminate the need for immunosuppressants. The decellularization process is performed using chemical detergents, enzyme degradation or irradiation [122]. Acellular nerve grafts are removed from Schwann cells and myelin, but the internal neuronal structure and extracellular matrix (collagen, laminin and growth factors) are preserved [123]. Regeneration process with acellular allografts involves host's migrated Schwann cells. Thus, even when acellular nerve grafts show good outcomes in trials, they are still insufficient in long nerve deficit repairs. In future, acellular grafts supplemented with seed cells, and growth factors may improve the surgical repair outcomes of large gap peripheral nerve injuries [124].

9.4. Nerve conduits

In recent years, research efforts have focused on the development of conduits as an alternative treatment especially for larger defects. Nerve conduits serve as a bridge between the proximal and distal stumps of the injured nerve and provide a scaffold for axonal regeneration. The most important advantage of a conduit is the ability to provide an ideal microenvironment for neuronal recovery. For this purpose, an ideal nerve conduit should have properties like biocompatibility, permeability, flexibility, biodegradability, compliance, neuroinductivity and neuroconductivity with appropriate surface [137, 138].

Conduits are categorized into two groups according to their materials as biological and synthetic conduits.

Biological nerve conduits include autologous arteries, veins, muscle, human amniotic membrane and umbilical cord vessels. Major advantages of biological conduits are non-activation of foreign body reaction, biocompatibility and enhanced migration of supportive cells. These biomaterials have been widely used for repair of short gap (<3 cm) nerve injuries, and the outcomes were consistent with those of nerve grafts [125–127, 138].

Synthetic nerve conduits include degradable and nondegradable conduits. Nondegradable nerve conduit materials include silicone, elastomeric hydrogel and porous stainless steel. Reconstruction with these materials is successful, but the possibility of foreign body reaction, scar tissue formation secondary to inflammation, lack of stability and the inflexible structure limit extensive use of them. Another disadvantage is the requirement of a second surgery for conduit removal.

Commonly used degradable conduit materials include collagen, polyesters (e.g., polyglycolic acid (PGA)), chitosan, polylactic acid (PLA) and hydrogel. These materials induce only minimal foreign body reaction, and several investigators reported effective nerve regeneration with these conduits [128]. Particularly, collagen conduits have shown comparable results with autologous nerve grafts in animal studies. There are many Food and Drug Administration (FDA) approved collagen-based conduits such as NeuraGen, NeuroFlex, NeuroMatrix, NeuroWrap and NeuroMend. Researchers observed that collagen conduits are resorbable, flexible, and cause minimal scar formation, allow nutrient transfer and provide suitable environment for nerve regeneration without any compression neuropathy. Over time, the material choice for nerve conduits shifted towards the use of more biocompatible synthetic polymers. Examples of these polymers are polylactic acid (PLA), polyglycolic acid (PGA), poly-caprolactone (PCL) and poly-lactide caprolactone (PLCL). Neurotube is a PGA nerve conduit, and Neurolac is a PLCL conduit. Neurotube was designed for gaps between 8 mm and 3 cm and is resorbed within 6–8 months. Neurolac is available as a tube with a length of 3 cm. In clinical trials, PGA conduits had comparable results with gold standards in the treatment of gaps up to 20 mm. A human study reported fistulization and neuroma in hand nerve surgery with PLCL conduits. Fibrin, gelatin, keratin and silk are other biopolymer conduit materials that are still under experimental evaluation [137, 138].

Studies have concluded that none of the experimental materials used for nerve conduits performed better than nerve grafts. However, observations showed that some of the properties make nerve conduits more useful (see **Table 1**). Conduits developed in the future should have a combination of aforementioned favourable properties.

9.5. Growth factors

Neurotrophic growth factors which are naturally released from injured nerves are used during nerve repair interventions. Studies about nerve repair pathophysiology reported that these growth factors act as cell regulators and they are all essential. Several studies showed that all these neurotrophins promote peripheral nerve regeneration [132, 133]. Additionally, neurotrophic growth factors promote both outgrowth and survival of motor and sensory neurons. Principal neurotrophic factors used for peripheral nerve regeneration are presented in **Table 2**. In experimental models, growth factors for nerve regeneration are generally used with nerve conduits.

9.6. Cell-based therapy

The limitations of present therapies are slow nerve regeneration and insufficient filling of large nerve gaps. To overcome these limitations, cell-based therapy was designed to provide supportive cells to the lesion site with the aim to accelerate nerve regeneration. Supportive

Should not harm peripheral tissue of the injury site
Resorbable
Limiting fibrous tissue infiltration
Allowing nutrient transfer with pores
Protein release (laminin/fibronectin)
Appropriate internal diameter to allow nerve regeneration
Release of neurotrophic factors
Allowing electrical conductance

Table 1. Desired features of the nerve conduits [137].

Neurotrophic factors	Effect
Nerve growth factor (NGF)	Survival signalling, neurite outgrowth
Glial cell line-derived neurotrophic factor (GDNF)	Sensory regeneration
Brain-derived neurotrophic factor (BDNF)	Positive modulation of peripheral nerve myelination
Neurotrophin-3 (NT-3)	Negative modulation of peripheral nerve myelination
Neurotrophin-4/5 (NT-4/5)	Survival of sensory neurons
Ciliary neurotrophic factors (CNTF)	Survival of motor neurons

Table 2. The effects of neurotrophic factors in peripheral nerve regeneration [137].

cell additions are combined with nerve grafts or conduits. Most extensively studied therapeutic models have included Schwann cells (SCs), but scientific improvements were achieved with different types of stem cells as well. Cell-based therapy is performed with stem cells owing to their self-renewal ability and capacity for differentiation into specialized cell types. Investigations on cell-based therapy are still in the preclinical level except some trials on Schwann cells [139]. Stem cells used for nerve repair interventions are mentioned in **Table 3**.

Schwann cells, bone marrow-derived mesenchymal stem cells (BMSCs), adipose-derived mesenchymal stem cells (ADSCs) and pluripotent stem cells are the primary cell types which are used in research studies. First of all, SCs are the most important and first-choice seed cells because they are the primary functional cells of the nervous system. SCs have a crucial role in nerve regeneration by producing neurotrophic factors such as nerve growth factor (NGF), brain-derived neurotrophic factor (BDNF), ciliary neurotrophic factor, platelet-derived growth factor and neuropeptide Y. In addition to being a supply for growth factors, SCs are capable of proliferation, immune modulation, remyelination and migration. All of these enhance repaired nerve healing. In cell-based therapies, neural crest cells are the main source of Schwann cells. SC seeds transplanted in a nerve conduit enhance axonal regeneration, but SC seeds have some disadvantages such as slow expansion to large numbers and being hard to obtain [129, 130, 140].

Embryonic stem cells (ESCs)
Neural stem cells (NSCs)
Bone marrow-derived stem cells (BMSCs)
Adipose-derived stem cells (ADSCs)
Skin-derived precursor stem cells (SKP-SCs)
Foetal-derived stem cells
Hair follicle stem cells (HFSCs)
Dental pulp stem cells (DPSCs)
Muscle-derived stem/progenitor cells (MDSPCs)
Induced pluripotential stem cells (iPSCs)

Table 3. Stem cells studied in peripheral nerve repair [139].

Embryonic stem cells (ESCs) have preferable advantages such as providing an unlimited source of cells, good differentiation potential and long-lasting proliferation capacity. However, ethical concerns are the major problem when these cells are used for transplantation. NSCs have the ability to differentiate into neurons and glial cells, but their use is limited because harvesting of these cells is difficult and there is a risk for formation of neuroblastoma. Bone marrow-derived stem cells (BMSCs) have the potential to differentiate into SC-like cells (BMSC-SCs). However, studies showed that differentiation potential of BMSCs is not as strong as NSCs. Also, harvesting procedure is painful and results in a lower cell fraction compared to other stem cells. Harvesting of ADSCs is minimally invasive and results in a high cellular yield. ADSCs can differentiate into SC-like cells and release BDNF, NGF and vascular endothelial growth factor (VEGF). These factors facilitate recruitment of endogenous SCs. The most important factor limiting the use of ADSCs is the potential to differentiate into adipocytes. Studies on BMSC-SCs and ADSC-SCs have yielded promising results [131, 140].

Foetal stem cells can be derived from amniotic fluid, amniotic membrane, umbilical cord and Wharton's jelly. Both amniotic tissue-derived stem cells (ATDSCs) and umbilical cord-derived mesenchymal stem cells (UC-MSCs) have differentiation and proliferation potential. ATDSCs also exhibit strong angiogenic potential and cause augmented neuronal injury perfusion. Tumorigenesis is one of the important side effects of UC-MSCs secondary to their high proliferation potential. Wharton's Jelly MSCs show specific mesenchymal features like generating neurotrophic factors such as NGF, BDNF and NT-3. Major advantages of foetal-derived stem cells are easy access and less immunoreactivity. Ethical issues are the most important disadvantage of foetal-derived stem cells. Skin-derived precursor stem cells (SKP-SCs) are found in the dermis and can differentiate into many kind of cells like neurons and glial cells. It is reported that SKP-SCs accelerate nerve regeneration. Hair follicle stem cells (HFSCs) have a unique feature of differentiation into SCs directly without any genetic intervention. Animal studies reported improved nerve repair with HFSCs. Several drawbacks associated with the stem cells have led to research efforts for alternative cells like induced pluripotential stem cells (iPSCs). iPSCs showed enhanced neuronal regeneration, but tumorigenicity, need for immunosuppression and chromosomal aberrations limit their use [139, 140].

As a result, ideal cells to be used for neural regeneration should have the properties of being suitable for easy harvesting, not requiring immunosuppression, being able to integrate to the injury site and being non-tumorigenic. And the success of a cell-based therapy depends on the transplanted cell's ability to differentiate into Schwann-like cells, to release neurotrophic growth factors and to induce myelinization of axons. Schwann cell cultures have mostly shown acceptable results in experimental studies; however, they are not good enough and search for an ideal cell is still ongoing. Bone marrow-derived mesenchymal cells have also demonstrated favourable results with numerous advantages like easy harvesting, high cell viability and secretion of multiple trophic factors [139, 140].

Even though cell-based therapy is promising for future, it is already associated with certain limitations. The most important issue is cell transplantation safety, and the other is that cell preparations are time consuming. These delays might cause the most appropriate intervention time for neuronal repair to be missed.

http://dx.doi.org/10.5772/intechopen.68345

9.7. Other methods

Freeze-thawed muscle graft, nerve transfer and direct muscular neurotization are the other surgical methods for nerve repair. In the nerve transfer surgery, the uninjured nerve is transferred to the distal stump of the injured nerve. During the direct muscular neurotization, the avulsed end of the nerve is implanted into the muscle directly. All of these techniques still need further improvement [133–136].

10. Conclusion

Microsurgical direct nerve repair is still gold standard for peripheral nerve repair whenever possible with a tension-free and early repair. It should not be forgotten that nerve repair that would give the optimum result requires healthy supportive tissue. If there is large nerve deficit, the autologous nerve graft is accepted as the gold standard. Second surgery, potential for neuroma formation and loss of donor nerve function are the main disadvantages of autologous nerve grafting. And all these limitations cause to perform acellular human nerve grafts. Acellular grafts have promising results but are still insufficient for large nerve deficits. At this stage, nerve conduits which stand out with the formation of an ideal microenvironment in the large nerve defects are increasingly foreground. But performing successful results with nerve conduits still requires combination of pharmacological and molecular therapies. Researches should focus on both axonal regeneration and achieving optimum microenvironment. In this context, it is important to manipulate Schwann cells. Because SCs provide supportive cell proliferation, remyelination and growth factor supply. Studies about performing SCs-like cells become more important because harvesting SCs is not an easy and safe procedure. BMSCs and ADSCs have promising results, but tumorigenicity, need for immunosuppression and long-time need for cell preparations limit widely accepted clinical use of stem cells. Recent studies focus on manipulation of SCs to improve their contribution to nerve regeneration by increased motility, ability of differentiation and producing neurotrophic factors.

In conclusion, the combination of genetically modified Schwann like cells and conduits or acellularized grafts as three-dimensional structured scaffolds will finally achieve optimum functional recovery after peripheral nerve injury.

Author details

Aysu Hayriye Tezcan

Address all correspondence to: aysndr@gmail.com

Anesthesiology and Reanimation Department, Kafkas University School of Medicine, Kars, Turkey

References

[1] Kouyoumdjian JA. Peripheral nerve injuries: A retrospective survey of 456 cases. Muscle & Nerve. 2006;**34**(6):785–788. DOI: 10.1002/mus.20624

[2] Siemionow M, Brzezicki G. Current techniques and concepts in peripheral nerve repair. International Review of Neurobiology. 2009;**87**:141–172. DOI: 10.1016/S0074-7742(09)87008-6

[3] Robinson MD, Bryant PR. Praxedes V. Belandres and Timothy R. Dillingham. Peripheral nerve injuries. Rehabilitation of the Injured Combatant. Vol. 2. 1st ed. Department of the Army; 1998. pp. 419–574

[4] Ryu J, Beimesch CF, Lalli TJ. (iii) Peripheral nerve repair. Orthopaedics Trauma. Washington. 2011;**25**: 174–118

[5] Goran Lundborg, The nerve trunk. Nerve Injury and Repair. 2nd ed. London: Churchill Livingstone; 1988. pp. 32–63

[6] Sunderland S. Nerves and Nerve Injuries. Edinburgh: Churchill Livingstone; 1978. p. 10

[7] Carp SJ. Peripheral Nerve Injury. An anatomical and Physiological Approach for Physical Therapy Intervention. 1st ed. Philadelphia: FA Davis Company; 2015. p. 1–14

[8] Seddon HJ. Three types of nerve injury. Brain. 1943;**66**:237-288. DOI: 10.1093/brain/66.4.237

[9] Menorca, RMG, Fussell TS, Elfar JC. Nerve physiology: Mechanisms of injury and recovery. Hand Clinics. 2013;**29**(3):317–330. DOI: 10.1016/j.hcl.2013.04.002

[10] Sunderland S. Traumatic injuries of peripheral nerves: 1. Simple compression injuries of the radial nerve. Brain. 1945;**68**:56–72. DOI: 10.1093/brain/68.1.56

[11] Burnett MG, Zager EL. Pathophysiology of peripheral nerve injury: A brief review. Neurosurgical Focus. 2004;**16**(5):1–7. DOI: 10.3171/foc.2004.16.5.2

[12] Waller A. Experiments on the section of the glossopharyngeal and hypoglossal nerves of the frog, and observations of the alterations produced thereby in the structure of their primitive fibers. Philosophical Transactions of the Royal Society of London. 1850;**140**:423–429

[13] Griffin JW, Hogan MV, Chhabra AB, Deal DN. Peripheral nerve repair and reconstruction. Journal of Bone and Joint Surgery. 2013;**95**(23):2144–2151. DOI: 10.2106/JBJS.L.00704

[14] Grinsell D, Keating CP. Peripheral nerve reconstruction after injury: A review of clinical and experimental therapies. BioMed Research International. 2014;**2014**:1–13. DOI: 10.1155/2014/698256

[15] Pfister BJ, Gordon T, Loverde JR, et al. Biomedical engineering strategies for peripheral nerve repair: Surgical applications, state of the art, and future challenges. Critical Reviews in Biomedical Engineering. 2011;**39**(2):81–124. DOI: 10.1615/CritRevBiomedEng.v39.i2.20

[16] Gölz G, Uhlmann L, Ludecke D, et al. The cytokine /neurotrophin axis in peripheral axon outgrowth. European Journal of Neuroscience. 2006;**24**:2721–2730. DOI: 10.1111/j.1460-9568.2006.05155.x

[17] Birch R. Nerve repair. In: Green DP, Hotchkiss RN, Pederson WC, Wolfe SW, editors. Green's Operative Hand Surgery. Philadelphia: Elsevier Churchill Livingstone; 2005. pp. 1075–1111

[18] Meeks GR, Gray JE. Up-to-date Nerve Injury Associated with Pelvic Surgery [Internet]. 2011. Available from: http://uptodate.com/contents/nerve-injury-associated-withpelvic-surgery

[19] Dillavou ED, Anderson LR, Bernert RA, et al. Lower extremity iatrogenic nerve injury due to compression during intraabdominal surgery. The American Journal of Surgery. 1997;**173**:504–508. DOI: 10.1016/S0002-9610(97)00015-9

[20] Saint-Arnaud D, Paquin MJ. Safe positioning for neurosurgical patients. AORN. 2008;**87**(6):1156–1168. DOI: 10.1016/j.aorn.2008.03.004

[21] Beckett AE. Are we doing enough to prevent patient injury caused by positioning for surgery? Journal of Perioperative Practice. 2010;**20**(1):26–29. PMID:20225718

[22] Kretschmer T, Heinen CW, Antoniadis G, Richter HP, Konig RW. Iatrogenic nerve injuries. Neurosurgery Clinics of North America. 2009;**20**(1):73–90. DOI: 10.1016/j.nec.2008.07.025

[23] Cooper A. Perioperative positioning injuries on the rise: What to do. OR Connect. 2011;**4**(3):21–29

[24] Uribe JS, Kolla J, Omar H, et al. Brachial plexus injury following spinal surgery. Journal of Neurosurgery Spine. 2010;**13**:552–558. DOI: 10.3171/2010.4.SPINE09682

[25] Warner MA, Marting JT, Schroeder DG, Offord KP, Chute CG. Lower-extremity motor neuropathy associated with surgery performed on patients in a lithotomy position. Anesthesiology. 1994;**81**(1):6–12. DOI: 10.1097/00000542-199407000-00004

[26] Parks BJ. Postoperative peripheral neuropathies. Surgery. 1973;**74**(3):348–357

[27] Sawyer RJ, Richmond MN, Hickey JD, Jarratt JA. Peripheral nerve injuries associated with anaesthesia. Anaesthesia. 2000;**55**:980–991. DOI: 10.1046/j.1365-2044.2000.01614.x

[28] Prielipp RC, Morell RC, Walker FO, Santos CC, Bennett J, Butterworth J. Ulnar nerve pressure. Anesthesiology. 1999;**91**(2):345–354. DOI: 10.1097/00000542-199908000-00006

[29] Akhavan A, Gainsburg DM, Stock JA. Complications associated with patient positioning in urologic surgery. Urology. 2010;**76**(6):1309–1316. DOI: 10.1016/j.urology.2010.02.060

[30] Barner KC, Landau ME, Campbell WW. A review of perioperative nerve injury to the lower extremities: Part 1. Journal of Clinical Neuromuscular Disease. 2002;**4**(2):95–99. DOI: 10.1097/00131402-200212000-00009

[31] Brown GD, Swanson EA, Nercessian OA. Neurologic injuries after total hip arthroplasty. American Journal of Orthopaedics. 2008;**37**(4):191–197. PMID:18535674

[32] Grocott HP, Clark JA, Homi HM, Sharma A. Other neurologic complications after cardiac surgery. Seminars in Cardiothoracic and Vascular Anesthesia. 2004;**8**(3):213–226. DOI: 10.1177/108925320400800304

[33] Pereles TR, Stuchin SA, Kastenbaum DM, Beric A, Lacagnino G, Kabir H. Surgical maneuvers placing the sciatic nerve at risk during total hip arthroplasty as assessed by somatosensory evoked potential monitoring. Journal of Arthroplasty. 1996;**11**(4):438–444. DOI: 10.1016/S0883-5403(96)80034-9

[34] Schwartz DM, Sestokas AK, Hilibrand AS, et al. Neurophysiological identification of position-induced neurologic injury during anterior cervical spine surgery. Journal of Clinical Monitoring and Computing. 2006;**20**:437–444. DOI: 10.1007/s10877-006-9032-1

[35] Bradshaw AD, Advincula AP. Postoperative neuropathy in gynecologic surgery. Obstetrics and Gynecology Clinics of North America. 2010;**37**:451–459. DOI: 10.1016/j.ogc.2010.05.008

[36] Kretschmer T, Antoniadis G, Braun V, et al. Evaluation of iatrogenic lesions in 722 surgically treated cases of peripheral nerve trauma. Journal of Neurosurgery. 2001;**94**(6):905–912. DOI: 10.3171/jns.2001.94.6.0905

[37] Welch MB, Brummett CM, Welch TD, et al. Perioperative peripheral nerve injuries: A retrospective study of 380,680 cases during a 10-year period at a single institution. Anesthesiology. 2009;**111**:490–497. DOI: 10.1097/ALN.0b013e3181af61cb

[38] Kim DH, Murovic JA, Tiel R, et al. Management and outcomes in 353 surgically treated sciatic nerve lesions. Journal of Neurosurgery. 2004;**101**(1):8–17. DOI: 10.3171/jns.2004.101.1.0008

[39] Kim DH, Kline DG. Surgical outcome for intra- and extrapelvic femoral nerve lesions. Journal of Neurosurgery. 1995;**83**(5):783–790. DOI: 10.3171/jns.1995.83.5.0783

[40] Kim DH, Cho YJ, Tiel RL, et al. Surgical outcomes of 111 spinal accessory nerve injuries. Neurosurgery. 2003;**53**(5):1106-1112. DOI: 10.3171/jns.1995.83.5.0783

[41] Antoniadis G, Kretschmer T, Pedro MT, et al. Iatrogenic nerve injuries. Deutsches Ärzteblatt International. 2014;**111**(16):273-279. DOI: 10.3238/arztebl.2014.0273

[42] Kretschmer T, Heinen CW, Antoniadis G, et al. Iatrogenic nerve injuries. Neurosurgery Clinics of North America. 2009;**20**(1):73-90. DOI: 10.1111/j.0042-9007.2004.572_6.x

[43] Kretschmer T, Antoniadis G, Börm W, et al. Iatrogenic nerve injuries. Der Chirurg. 2004;**75**(11):1104–1112. DOI: 10.1016/j.nec.2008.07.025

[44] McGeorge D, Sturzenegger M, Buchler U. Tibial nerve mistakenly used as a tendon graft. Reports of three cases. The Bone and Joint Journal. 1992;**74**(3):365–366. DOI: PMID: 1587878

[45] Geldmacher J. Median nerve as free tendon graft. Journal of Hand Surgery. 1972;**4**(1):56. DOI: 10.1016/0072-968X(72)90012-5

[46] Weber RV, Mackinnon SE. Median nerve mistaken for palmaris longus tendon: Restoration of function with sensory nerve transfers. Hand. 2007;**2**:1-4. DOI: 10.1007/s11552-006-9011-5

[47] Oppikofer C, Tschopp H. Tibial nerve used as a tendon graft. Helvetica Chirurgica Acta. 1990;**57**:923–929

[48] Kim DH, Murovic JA, Tiel RL, Kline DG. Operative outcomes of 546 Louisiana State University Health Sciences Center peripheral nerve tumors. Neurosurgery Clinics of North America. 2004;**15**:177–192. DOI: 10.1016/j.nec.2004.02.006

[49] Lalkhen AG, Bhatia K. Perioperative peripheral nerve injuries. Continuing Education in Anaesthesia Critical Care & Pain. 2012;**12**(1):38–42. DOI: 10.1093/bjaceaccp/mkr048

[50] Laxton CH, Kipling R. Lingual nerve paralysis following the use of the laryngeal mask airway. Anaesthesia. 1996;**51**:869–870. DOI: 10.1111/j.1365-2044.1996.tb12621.x

[51] Brull R, Hafzic A, Reina MA, et al. Pathophysiology and etiology of nerve injury following peripheral nerve blockade. Regional Anesthesia and Pain Medicine. 2015;**40**(5):479–490. DOI: 10.1097/AAP.0000000000000125

[52] Hogan QH. Pathophysiology of peripheral nerve injury during regional anesthesia. Regional Anesthesia and Pain Medicine. 2008;**33**:435–441. DOI: 10.1097/00115550-200809000-00006

[53] Gentili F, Hudson A, Kline DG, et al. Peripheral nerve injection injury: An experimental study. Neurosurgery. 1979;**4**:244–253. DOI: 10.1227/00006123-197903000-00009

[54] Partridge BL. The effects of local anesthetics and epinephrine on rat sciatic nerve blood flow. Anesthesiology. 1991;**75**:243–250. DOI: 10.1097/00000542-199108000-00012

[55] Rodriguez J, Taboada M, Garcia F, et al. Intraneural hematoma after nerve stimulation–guided femoral block in a patient with factor XI deficiency: Case report. Journal of Clinical Anesthesia. 2011;**23**:234–237. DOI: 10.1016/j.jclinane.2010.03.009

[56] Steinfeldt T, Werner T, Nimphius W, et al. Histological analysis after peripheral nerve puncture with pencil-point or Tuohy needletip. Anesthesia and Analgesia. 2011;**112**:465–470. DOI: 10.1213/ANE.0b013e318202cb9c

[57] Moen V, Dahlgren N, Irestedt L. Severe neurological complications after central neuraxial blockades in Sweden 1990–1999. Anesthesiology. 2004;**101**:950–959. DOI: 10.1097/00000542-200410000-00021

[58] Neal JM, Barrington MJ, Brull R, et al. The second ASRA practice advisory on neurologic complications associated with regional anesthesia and pain medicine: Executive summary 2015. Regional Anesthesia and Pain Medicine. 2015;**40**(5):401–430. DOI: 10.1097/AAP.0000000000000286

[59] American Society of Anesthesiologists Task Force on Prevention of Perioperative Peripheral Neuropathies. Practice advisory for the prevention of perioperative peripheral

neuropathies: An updated report by the American Society of Anesthesiologists Task Force on prevention of perioperative peripheral neuropathies. Anesthesiology. 2011;**114**(4):741. DOI: 10.1097/ALN.0b013e3181fcbff3

[60] Horlocker TT, Hebl JR, Gali B, et al. Anesthetic, patient, and surgical risk factors for neurologic complications after prolonged total tourniquet time during total knee arthroplasty. Anesthesia and Analgesia. 2006;**102**(3):950–955. DOI: 10.1213/01.ane.0000194875.05587.7e

[61] Lundborg G, Dahlin LB. The pathophysiology of nerve compression. Hand Clinics. 1992;**8**(2):215–227. PMID:1613031

[62] Murphy CG, Winter DC, Bouchier-Hayes DC. Tourniquet injuries: Pathogenesis and modalities for attenuation. Acta Orthopaedica Belgica. 2005;**71**(6):635–645. PMID:1645 9851

[63] Odinsson A, Finsen V. Tourniquet use and its complications in Norway. Journal of Bone and Joint Surgery. 2006;**88**(8):1090–1092. DOI: 10.1302/0301-620X.88B8.17668

[64] Sharma JP, Salhotra R. Tourniquets in orthopedic surgery. Indian Journal of Orthopaedics. 2012;**46**(44):377–383. DOI: 10.4103/0019-5413.98824

[65] Caissie R, Goulet J, Fortin M, et al. Iatrogenic paresthesia in the third division of the trigeminal nerve: 12 years of clinical experience. Journal of the Canadian Dental Association. 2005;**71**(3):185–190. PMID:15763037

[66] Zuniga JR, LaBanc JP. Advances in microsurgical nerve repair. Journal of Oral Maxillofacial Surgery. 1993;**51**(1):62–68. DOI: 10.1016/0278-2391(93)90010-B

[67] Pogrel MA. The results of microneurosurgery of the inferior and lingual nerve. Journal of Oral Maxillofacial Surgery. 2002;**60**(5):485–489. DOI: 10.1053/joms.2002.31841

[68] Ardito G, Revelli L, D'Alatri L, et al. Revisited anatomy of the recurrent laryngeal nerves. The American Journal of Surgery. 2004;**187**(2):249–253. DOI: 10.1016/j.amjsurg.2003. 11.001

[69] Tewari A, Samy RN, Castle J, et al. Intraoperative neurophysiological monitoring of the laryngeal nerves during anterior neck surgery: A review. Annals of Otology, Rhinology & Laryngology. 2017;**126**(1):67–72. DOI: 10.1177/0003489416675354

[70] Farrar W. Complications of thyroidectomy. Surgery Clinics of North America. 1983;**63**:1353-1361. DOI: 10.1016/S0039-6109(16)43193-2

[71] Lipton RJ, McCafferey TV, Litchy WJ. Intraoperative electrophysiologic monitoring of laryngeal muscle during thyroid surgery. Laryngoscope. 1988;**98**:1292–1296. DOI: 10.1288/00005537-198812000-00003

[72] Patlow C, Norton J, Brennan MF. Vocal cord paralysis and reoperative parathyroidectomy. Annals of Surgery. 1986;**203**:282–285. DOI: 10.1097/00000658-198603000-00011

[73] Holt GR, McMurray GR, Joseph DL. Recurrent laryngeal nerve injury following thyroid operations. Surgery Gynecololgy and Obstetrics. 1977;**4**:567–570. PMID:847613

[74] Jung A, Schramm J, Lehnerdt K, Herberhold C. Recurrent laryngeal nerve palsy during anterior cervical spine surgery: A prospective study. Journal of Neurosurgery Spine. 2005;**2**:123–127. DOI: 10.3171/spi.2005.2.2.0123

[75] Filho JOG, Kowalski LP. Surgical complications after thyroid surgery performed in a cancer hospital. Otolaryngology Head and Neck Surgery. 2005;**132**(3):490–494. DOI: 10.1016/j.otohns.2004.09.028

[76] Pezzulo L, Dalrio P, Losito NS, et al. Post-operative complications after completion thyroidectomy for differentiated thyroid cancer. European Journal of Surgical Oncology. 1997;**123**:215–218. DOI: 10.1016/S0748-7983(97)92340-4

[77] Jiang Y, Gao B, Zhang X, et al. Prevention and treatment of recurrent laryngeal nerve injury in thyroid surgery. International Journal of Clinical and Experimental Medicine. 2014;**7**(1):101–107. PMID:24482694

[78] Wang W, Chen D, Chen S, et al. Laryngeal reinnervation using ansa cervicalis for thyroid surgery-related unilateral vocal fold paralysis: A long-term outcome analysis of 237 cases. PLoS One. 2011;**6**:e19128. DOI: 10.1371/journal.pone.0019128

[79] Yumoto E, Sanuki T, Toya Y, et al. Nerve-muscle pedicle flap implantation combined with arytenoids adduction. Archives of Otolaryngology Head Neck Surgery. 2010;**136**:965–969. DOI: 10.1001/archoto.2010.155

[80] Sulica L. The superior laryngeal nerve: Function and dysfunction. Otolaryngology Clinics North America. 2004;**37**:183–201. DOI: 10.1016/S0030-6665(03)00175-0

[81] Nasseri SS, Maragos NE. Combination thyroplasty and the 'twisted larynx:' Combined type IV and type I thyroplasty for superior laryngeal nerve weakness. Journal of Voice. 2000;**14**:104–111. DOI: 10.1016/S0892-1997(00)80100-9

[82] Shaw G, Searl J, Hoover L. Diagnosis and treatment of unilateral cricothyroid muscle paralysis with a modified Isshiki type 4 thyroplasty. Otolaryngology Head Neck Surgery. 1995;**113**:679–688. DOI: 10.1016/S0194-5998(95)70005-6

[83] El-Kashlan HK, Carroll WR, Hogikyan MD, et al. Selective cricothyroid muscle reinnervation by muscle-nerve-muscle neurotization. Archives of Otolaryngology Head Neck Surgery. 2001;**127**:1211–1215. DOI: 10.1001/archotol.127.10.1211

[84] Hall MB, Brown RW, Lebowitz MS. Facial nerve injury during surgery of the temporomandibular joint: A comparison of two dissection techniques. Journal of Oral Maxillar Surgery. 1985;**43**:20–23. DOI: 10.1016/S0278-2391(85)80008-2

[85] Wilson L, Lin E, Lalwani A. Cost-effectiveness of intraoperative facial nerve monitoring in middle ear or mastoid surgery. Laryngoscope. 2003;**113**:1736–1745. DOI: 10.1097/00005537-200310000-00015

[86] Daube JR, Harner SG, Ebersold MJ. Electrophysiologic monitoring of facial nerve during temporal bone surgery. Laryngoscope. 1986;**96**:65–69. DOI: 10.1288/00005537-19860 1000-00012

[87] Hohman MH, Bhama PK, Hadlock TA. Epidemiology of iatrogenic facial nerve injury: A decade of experience. The Laryngoscope. 2014;**124**(1):260–265. DOI: 10.1002/lary.24117

[88] Frijters E, Hofer SO, Mureau MA. Long-term subjective and objective outcome after primary repair of traumatic facial nerve injuries. Head Neck Surgery. 2008;**61**(2):181–187. DOI: 10.1097/SAP.0b013e3181591e27

[89] Malik TH, Kelly G, Ahmed A, et al. A comparison of surgical techniques used in dynamic reanimation of the paralyzed face. Otology & Neurotology. 2005;**26**(2):284–291. DOI: 10.1097/00129492-200503000-00028

[90] Boström D, Dahlin LB. Iatrogenic injury to the accessory nerve. Scandinavian Journal of Plastic and Reconstructive Surgery and Hand Surgery. 2007;**41**:82–87. DOI: 10.1080/02844310600836810

[91] Bradley PJ, Ferlito A, Silver CE, et al. Neck treatment and shoulder morbidity: Still a challenge. Head Neck. 2011;**33**:1060–1067. DOI: 10.1002/hed.21495

[92] Bäzner UM, Braun V, Richter HP, et al. Management iatrogener N.-accessorius-Läsionen. Der Nervenarzt. 2005;**76**(4):462–466. DOI: 10.1007/s00115-004-1796-2

[93] Sharma AD, Parmley CL, Sreeram G, et al. Peripheral nerve injuries during cardiac surgery: Risk factors, diagnosis, prognosis, and prevention. Anesthesia Analgesia. 2000;**91**:1358. DOI: 10.1097/00000539-200012000-00010

[94] DeVita MA, Robinson LR, Rehder J, et al. Incidence and natural history of phrenic neuropathy occurring during open heart surgery. Chest. 1993;**103**:850–856. DOI: 10.1378/chest.103.3.850

[95] Laub GW, Muralidharan S, Chen C, et al. Phrenic nerve injury. A prospective study. Chest. 1991;**100**:376–379. DOI: 10.1378/chest.101.5.1479-b

[96] Nahabedian MY, Dellon AL. Outcome of the operative management of nerve injuries in the ilioinguinal region. Journal of the American College of Surgeons. 1997;**184**(3):265268. PMID: 9060923

[97] Muto CM, Pedana N, Scarpelli S, et al. Inguinal neurectomy for nerve entrapment after open/laparoscopic hernia repair using retroperitoneal endoscopic approach. Surgical Endoscopy and Other Interventional Techniques. 2005;**19**(7):974–976. DOI: 10.1007/s00464-004-2007-8

[98] Krahenbuhl L, Striffeler H, Baer HU, et al. Retroperitoneal endoscopic neurectomy for nerve entrapment after hernia repair. British Journal of Surgery. 1997;**84**(2):216–219. DOI: 10.1046/j.1365-2168.1997.02629.x

[99] Grosz CR. Iliohypogastric nerve injury. American Journal of Surgery. 1981;**142**:628. DOI: 10.1016/0002-9610(81)90441-4

[100] Kim DH, Murovic JA, Tiel RL, et al. Surgical management of 33 ilioinguinal and iliohypogastric neuralgias at Louisiana State University Health Sciences Center. Neurosurgery. 2005;**56**:1013–1020. PMID: 15854249

[101] Immerman I, Price AE, Alfonso I, et al. Lower extremity nerve trauma. Bulletin of the Hospital for Joint Diseases. 2014;**72**(1):43–52.

[102] Irvin W, Andersen W, Taylor P, et al. Minimizing the risk of neurologic injury in gynecologic surgery. Obstetrics Gynecology. 2004;**103**:374. DOI: 10.1097/01.AOG.0000110542.53489.c6

[103] Kim DH, Murovic JA, Tiel RL, et al. Intrapelvic and thigh-level femoral nerve lesions: Management and outcomes in 119 surgically treated cases. Journal of Neurosurgery. 2004;**100**:989-996. DOI: 10.3171/jns.2004.100.6.0989

[104] Schmalzried TP, Noordin S, Amstutz HC. Update on nerve palsy associated with total hip replacement. Clinical Orthopaedics and Related Research. 1997;**344**:188–206. DOI: 10.1097/00003086-199711000-00020

[105] Farrell CM, Springer BD, Haidukewych GJ, et al. Motor nerve palsy following primary total hip arthroplasty. Journal of Bone and Joint Surgery. 2005;**87**(12):2619–2625. DOI: 10.2106/00004623-200512000-00004

[106] Yeremeyeva E, Kline DG, Kim DH. Iatrogenic sciatic nerve injuries at buttock and thigh levels: The Louisiana State University experience review. Neurosurgery. 2009;**65**(4):A63-A66. DOI: 10.1227/01.NEU.0000346265.17661.1E

[107] Telleria JJ, Safran MR, Harris AH, et al. Risk of sciatic nerve traction injury during hip arthroscopy-is it the amount or duration? An intraoperative nerve monitoring study. Journal of Bone and Joint Surgery. 2012;**21**:2025–2032. DOI: 10.2106/JBJS.K.01597

[108] Aldea PA, Shaw WW. Lower extremity nerve injuries. Clinics in Plastic Surgery. 1986;**13**(4):691–699. PMID: 3533378

[109] Wildner M, Peters A, Hellich J, et al. Complications of high tibial osteotomy and internal fixation with staples. Archives in Orthopaedics Trauma Surgery. 1992;**111**(4):210–212. DOI: 10.1007/BF00571479

[110] Asp JPL, Rand JA. Peroneal nerve palsy after total knee arthroplasty. Clinics in Orthopaedics Related Research. 1990;**261**:233–237. PMID:2245551

[111] Idusuyi OB, Morrey BF. Peroneal nerve palsy after total knee arthroplasty. Assessment of predisposing and prognostic factors. Journal of Bone and Joint Surgery. 1996;**78**(2):177–184. PMID:8609107

[112] Schinsky MF, Macaulay W, Parks ML, et al. Nerve injury after primary total knee arthroplasty. Journal of Arthroplasty. 2001;**16**(8):1048–1054. DOI: 10.1054/arth.2001.26591

[113] Nercessian OA, Ugwonali OF, Park S. Peroneal nerve palsy after total knee arthroplasty. Journal of Arthroplasty. 2005;**20**(8):1068–1073. DOI: 10.1016/j.arth.2005.02.010

[114] Tandon PN, Ramamurthi R, Jain PK. Ramamurthi & Tandon's Manual of Neurosurgery. 1st ed. New Delhi, India: Jaypee Brothers Medical Publishers (P) Ltd.; 2014. p. 409–415

[115] Winn HR. Youmans Neurological Surgery. 6th ed. Atlanta, USA: Elsevier Sounders; 2011. p. 2388–2397.

[116] Lee SK, Wolfe SW. Peripheral nerve injury and repair. Journal of the American Academy of Orthopaedic Surgeons. 2000;**8**(4):243–252. PMID:10951113

[117] Sameem M, Wood TJ, Bain JR. A systematic review on the use of fibrin glue for peripheral nerve repair. Plastic and Reconstructive Surgery. 2011;**127**(6):2381–2390. DOI: 10.1097/PRS.0b013e3182131cf5

[118] Gu X, Ding F, Yang Y, Liu J. Construction of tissue engineered nerve grafts and their application in peripheral nerve regeneration. Progress in Neurobiology. 2011;**93**(2):204–230. DOI: 10.1016/j.pneurobio.2010.11.002

[119] Nichols CM, Brenner MJ, Fox IK, et al. Effects of motor versus sensory nerve grafts on peripheral nerve regeneration. Experimental Neurology. 2004;**190**(2):347–355. DOI: 10.1016/j.expneurol.2004.08.003

[120] Panseri S, Cunha C, Lowery J, et al. Electrospun micro- and nanofiber tubes for functional nervous regeneration in sciatic nerve transections. BMC Biotechnology. 2008;**8**(1):39. DOI: 10.1186/1472-6750-8-39

[121] Mackinnon SE, Doolabh VB, Novak CB, et al. Clinical outcome following nerve allograft transplantation. Plastic and Reconstructive Surgery. 2001;**107**(6):1419–1429. DOI: 10.1097/00006534-200105000-00016

[122] Karabekmez FE, Duymaz A, Moran SL. Early clinical outcomes with the use of decellularized nerve allograft for repair of sensory defects within the hand. Hand. 2009;**4**(3):245–249. DOI: 10.1007/s11552-009-9195-6

[123] Zhang Y, Luo H, Zhang Z, et al. A nerve graft constructed with xenogeneic acellular nerve matrix and autologous adipose-derived mesenchymal stem cells. Biomaterials. 2010;**31**(20):5312–5324. DOI: 10.1016/j.biomaterials.2010.03.029

[124] Nadim W, Anderson PN, Turmaine M. The role of Schwann cells and basal lamina tubes in the regeneration of axons through lengths of freeze-killed nerve grafts. Neuropathology and Applied Neurobiology. 1990;**16**(5):411–421. DOI: 10.1111/j.1365-2990.1990.tb01277.x

[125] Schmidt CE, Leach JB. Neural tissue engineering: strategies for repair and regeneration. Annual Review of Biomedical Engineering. 2003;**5**:293–347. DOI: 10.1146/annurev.bioeng.5.011303.120731

[126] Chiu DWT, Strauch B. A prospective clinical evaluation of autogenous vein grafts used as a nerve conduit for distal sensory nerve defects of 3 cm or less. Plastic and Reconstructive Surgery. 1990;**86**(5):928–934. PMID:2236318

[127] Archibald SJ, Shefner J, Krarup C, et al. Monkey median nerve repaired by nerve graft or collagen nerve guide tube. Journal of Neuroscience. 1995;**15**(5pt2):4109–4123. PMID:7751969

[128] Tohill M, Mantovani C, Wiberg M, Terenghi G. Rat bone marrow mesenchymal stem cells glial markers and stimulate nerve regeneration. Neuroscience Letters. 2004;**362**(3):200–203. DOI: 10.1016/j.neulet.2004.03.077

[129] Radtke C, Aizer AA, Agulian SK, et al. Transplantation of olfactory ensheathing cells enhances peripheral nerve regeneration after microsurgical nerve repair. Brain Research. 2009;**13**(1254):10–17. DOI: 10.1016/j.brainres.2008.11.036

[130] Mingo-Robinet J, Castañeda-Cabrero C, Alvarez V, et al. Tourniquet-related iatrogenic femoral nerve palsy after knee surgery: Case report and review of the literature. Case Reports in Orthopedics. 2013;**2013**:1–4. DOI: 10.1155/2013/368290

[131] Ohta M, Suzuki Y, Chou H, et al. Novel heparin/alginate gel combined with basic fibroblast growth factor promotes nerve regeneration in rat sciatic nerve. Journal of Biomedical Materials Research Part A. 2004;**71**(4):661–668. DOI: 10.1002/jbm.a.30194

[132] Wang S, Cai Q, Hou J, et al. Acceleration effect of basic fibroblast growth factor on the regeneration of peripheral nerve through a 15-mm gap. Journal of Biomedical Materials Research Part A. 2003;**66**(3):522–531. DOI: 10.1002/jbm.a.10008

[133] Lawson GM, Glasby MA. A comparison of immediate and delayed nerve repair using autologous freeze thawed muscle grafts in a large experimental animal model. The simple injury. Journal of Hand Surgery. 1995;**20**:663e70. DOI: 10.1016/S0266-7681(05)80131-7

[134] Ozkan T, Ozer K, Gulgonen A. Restoration of sensibility in irreparable ulnar and median nerve lesions with use of sensory nerve transfer: Long-term follow up of 20 cases. Journal of Hand Surgery. 2001;**26**:44e51. DOI: 10.1053/jhsu.2001.20156

[135] Brunelli G, Monini L. Direct muscular neurotization. Proceedings of the second congress of the international federation of societies for surgery of the hand. Journal of Hand Surgery. 1985;**10**:993e4. DOI: 10.1016/S0363-5023(85)80022-8

[136] Arslantunali D, Dursun T, Yucel D, et al. Peripheral nerve conduits: Technology update. Medical Devices. 2014;**7**:405. DOI: 10.2147/MDER.S59124

[137] Nectow AR, Marra KG, Kaplan DL. Biomaterials for the development of peripheral nerve guidance conduits. Tissue Engineering Part B: Reviews. 2011;**18**(1):40–50. DOI: 10.1089/ten.teb.2011.0240

[138] Muheremu A, Ao Q. Past, present and future of nerve conduits in the treatment of peripheral nerve injury. BioMed Research International. 2015;**2015**:1–6. DOI: 10.1155/2015/237507

[139] Jiang L, Jones S, Jia X. Stem cell transplantation for peripheral nerve regeneration: Current Options and opportunities. 2017;**18**:94–111. DOI: 1 0.3390/ijms18010094

[140] Sullivan R, Dailey T, Duncan K, et al. Peripheral nerve injury: Stem cell therapy and peripheral nerve transfer. International Journal of Molecular Sciences. 2016;**17**:2101–2112. DOI: 1 0.3390/ijms17122101

Chapter 2

Peripheral Nerve Entrapment and their Surgical Treatment

Vicente Vanaclocha-Vanaclocha,
Nieves Sáiz-Sapena, Jose María Ortiz-Criado and
Nieves Vanaclocha

Additional information is available at the end of the chapter

http://dx.doi.org/10.5772/67946

Abstract

Nerves pass from one body area to another through channels made of connective tissue and/or bone. In these narrow passages, they can get trapped due to anatomic abnormalities, ganglion cysts, muscle or connective tissue hypertrophy, tumours, trauma or iatrogenic mishaps. Nearly all nerves can be affected. The clinical presentation is pain, paraesthesia, sensory and motor power loss. The specific clinical features will depend on the affected nerve and on the chronicity, severity, speed and mechanism of compression. Its incidence is higher under some occupations and is some systemic conditions: diabetes mellitus, hypothyroidism, acromegaly, alcoholism, oedema and inflammatory diseases. The diagnosis is suspected with the clinical presentation and provocative clinical test, being confirmed with electrodiagnostic and/or ultrasonographic studies. Magnetic Resonance Studies (MRI) rule out ganglion cysts or tumours. Conservative medical treatment is often sufficient. In refractory ones, surgical decompression should be performed before nerve damage and muscle atrophy are irreversible. The 'double crash' syndrome happens when a peripheral nerve is compressed at more than one point along its trajectory. In cases with marked muscle atrophy, a 'supercharge end-to-side' nerve transfer can be added to the decompression. After decompression in those few cases with refractory pain, a nerve neurostimulator can be applied.

Keywords: entrapment neuropathy, compression neuropathy, carpal tunnel syndrome, cubital tunnel syndrome, meralgia paraesthetica, cheiralgia paraesthetica, peroneal nerve entrapment, ulnar tunnel syndrome, radial tunnel syndrome, tarsal tunnel syndrome

1. Introduction

Nerves pass from one body area or cavity to another through holes and channels made of connective tissue and/or a bone channel, be it total (mental nerve) or partial (carpal tunnel). In these narrow passages, they can get trapped and/or injured due to congenital anatomical abnormalities, muscle or connective tissue hypertrophy, ganglion cysts, tumours, trauma or iatrogenic mishaps. Nearly all nerves can suffer an entrapment syndrome. Not all have the same incidence, some being very common (i.e. carpal tunnel) [1] and some exceedingly rare (i.e. tarsal tunnel) [2].

Entrapment neuropathies results in pain, paraesthesia and muscle power loss in the distribution of a peripheral nerve. With time muscle atrophy and skin trophic changes will appear. The clinical presentation will depend on the specific affected nerve, the chronicity, severity, speed and mechanism of compression [3, 4].

Nerve entrapment incidence is higher under some systemic conditions: diabetes mellitus, hypothyroidism, acromegaly, chronic alcoholism, extensive oedema and systemic inflammatory diseases [4]. Some occupations are associated with specific peripheral nerve entrapment syndromes. For example, occupations requiring repetitive wrist or finger movements or handling of vibrating tools have a higher incidence of carpal tunnel syndrome (CTS) [5–7].

Clinical presentation and provocative tests will suggest a diagnosis [3] confirmed or not with electrodiagnostic or ultrasonographic studies [4, 8]. Moreover, electrodiagnostic studies are also helpful to stage the severity and to rule out other confounding conditions (i.e. carpal tunnel and C_7 radiculopathy, peroneal nerve compression vs. L_5 radiculopathy) [3, 9] or generalized diseases (i.e. diabetic peripheral neuropathy) [9, 10]. MRI studies often show changes, ganglion cysts or tumours [11, 12] but the ultrasonography is less costly and more easily available [13].

Conservative treatment is sufficient in many cases (i.e. Saturday night palsy) but otherwise surgical decompression should be considered before irreversible peripheral nerve damage and muscle atrophy are established [3, 4, 8].

A nerve can be compressed at more than one single point, exacerbating the effects [14, 15]. This is called the '*double crash*' syndrome and is common in some systemic diseases, particularly in diabetes mellitus [16].

In cases with advanced muscle atrophy, a *'supercharge end-to-side' nerve* transfer is an option. After thorough decompression, a nearby healthy nerve is sectioned and sutured to the side of the previously compressed nerve, ideally distal to the entrapment point. The motor axons of the healthy nerve will grow inside of the damaged one much faster than the damaged axons of the damaged nerve. So, the healthy axons of the healthy nerve will keep the muscle alive while the axons of the damaged nerve recover [4]. This has been performed between the *pronator quadratus* (PQ) nerve branch and the motor fascicle of the ulnar nerve (UN) at the forearm, between the flexor *digitorum superficialis* (FDS) and the anterior *interosseous* syndrome (AIN), between the triceps long head branch of the radial nerve (RN) and the axillary nerve, between the medial pectoral nerve and the axillary or the musculocutaneous nerve and between the spinal accessory nerve and the suprascapular nerve [4].

CTS is the most frequent entrapment syndrome, followed by meralgia paraesthetica and UN in the elbow. Decompression is always the treatment, removing the fibrous band, muscle or benign lesion causing the entrapment. After decompression, cases with refractory pain can undergo a nerve neurostimulator to block the pain transmission.

2. Upper extremity entrapment syndromes

2.1. Carpal tunnel syndrome (CTS)

This tunnel is formed by the 'U' of the carpal bones closed by the transverse carpal ligament. It is the most frequent entrapment neuropathy and one of the most common surgical conditions [17, 18]. Its estimated prevalence is 2% in men and 3% in women [17, 18], affecting a 3.72% of the USA population [1].

Idiopathic forms are due to a connective tissue proliferation of the flexor tendons synovium [19]. Some medical conditions predispose to its development: diabetes mellitus [20, 21], acromegaly [22], obesity [23], pregnancy [24], amyloidosis [25], hypothyroidism [22], rheumatoid arthritis [26], chronic kidney disease [27] and haemodialysis [28]. Its incidence is higher in occupations requiring repetitive finger and wrist movements [29], handling of vibrating tools [5–7] or repetitive blows with the palm of the hand (carpenters, sculptors) [29], but not with keyboard use [30]. It affects 30% of diabetics with polyneuropathy and 14% without it [31]. In pregnancy, it is most common in the third trimester [32].

Patients notice pain, numbness and tingling in the first three fingers of the hand. Initially symptoms are intermittent but become permanent with time, worsening with activity and at night [3]. Symptoms wake patients up at dawn, making them shake the affected hand to get rid of the symptoms (the so called *flick sign*). Paraesthesia may affect the whole hand. The pain may be an early symptom and radiate to the forearm or even to the whole arm up to the shoulder [33].

Sensory deficits affect the thumb, index and middle fingers and spare the thenar eminence [34], but 20% of clear-cut CTS show no sensory abnormalities [35]. Because the palmar cutaneous branch for the thenar eminence branches off the MN a few centimetres before the carpal tunnel the sensation of this area is normal in CTS. If this sensation is impaired pre-operatively it indicates proximal MN compression [3] while if damaged is only seen post-operatively it indicates iatrogenic injury.

Entrapment of this branch is possible but exceedingly rare [36].

Atrophy of the thenar muscles is a very late event in the progression of the disease [3] (**Figure 1C**), as are motor symptoms (**Figure 1D**) such as hand clumsiness, rigidity and loss of dexterity.

Symptoms are usually bilateral but predominate in one hand.

The diagnosis is suspected by the symptoms and provocative manoeuvres (Phalen test (**Figure 1A**) and the Tinel and the carpal compression signs) [37]. The Phalen test indicates advanced disease [38], having a 75% sensitivity and a 47% specificity [39]. Electrodiagnostic studies confirm the diagnosis, rule out confounding conditions and stage the disease [40], with an 85% sensitivity and a 95% specificity [41]. Symptoms do not always correlate with electrodiagnostic findings. Ultrasonography is also useful [42]. Due to its higher costs, MRI is not used regularly [43].

Up to 20% of CTS cases improve with conservative treatments [44, 45]. Night-time wrist splints help 60% of patients but many eventually need an operation [46, 47]. Local corticosteroid injections can provide relief but often temporary [48]. Surgical decompression is the only proven long-term lasting relief [40, 49]. Any concomitant systemic disease predisposing to CTS should be treated at once although decompression is usually needed nonetheless [27]. The surgical procedure entails complete transverse carpal ligament section to decompress the MN. Local, regional or general anaesthesia are options, but local is faster and more cost effective [50, 51]. Open field (**Figures 1F** and **G**) or endoscopy has a similar time out of work, but the latter MN damage is more frequent [52–54]. Retinaculotome decompression is similar to endoscopy but with less time and cost requirements [55] (**Figures 1H** and **I**). Re-operation is indicated in failure or recurrence. Incomplete decompression either at the distal carpal ligament or at the proximal antebrachial fascia is a frequent finding [56], but sometimes there is a thick scar tissue recreating the transverse carpal ligament and fixing the MN [4].

2.2. Pronator *teres* syndrome (PTS)

It is the MN compression as it passes through the pronator *teres* muscle (PTM) [57], the proximal arch of the FDS, the bicipital aponeurosis, the ligament of Struthers or an accessory head of the flexor *pollicis* longus muscle (FPL) (Gantzer's muscle) [58]. There is proximal forearm deep pain with sensory and/or motor deficits in the distribution of distal MN [33, 57]. Repetitive pronation aggravates symptoms, and contrary to CTS, they appear at daytime and disappear during the night. Another difference is that in PTS there can be sensory loss in the thenar eminence [58]. The pronator compression test is a steady digital pressure on the proximal edge of the PTM 6 cm distal to the elbow crease and 4 cm lateral to the medial epicondyle for 30 s [57]. If positive, it should reproduce the symptoms. Forearm pain and hand MN paraesthesias can be induced by resistive pronation or by elbow extension with the forearm in pronation [57], elbow flexion with the arm supinated or flexion of the middle finger interphalangeal joint [58]. Electrodiagnostic studies can confirm the diagnosis [59]. Conservative treatment should be attempted, encouraging patients to avoid pronation, particularly against resistance. If symptoms persist, surgical decompression is recommended [60]. It is usually performed open field but some have reported an endoscopic approach [57]. The number of cases is limited, so no definitive conclusions on the best technique can be drawn yet.

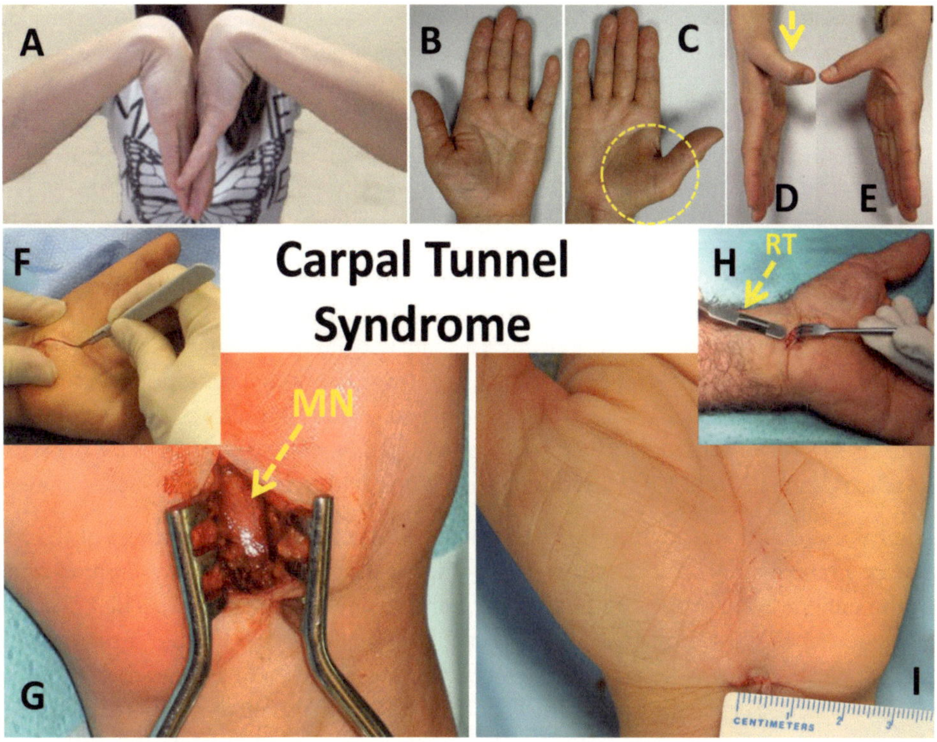

Figure 1. Carpal tunnel syndrome. (A) Phalen sign; (B) normal hand; (C) atrophy thenar eminence; (D) testing for weakness in motor thenar branch of median nerve; (E) normal hand; (F) and (G) open carpal tunnel release, median nerve (MN); (H) and (I) carpal tunnel decompression with retinaculotome (RT) or with endoscope.

2.3. Anterior interosseous syndrome

It is due to compression of this purely motor branch of the MN [61]. It induces a mild vague forearm pain accompanied by paresis or complete paralysis of the FPL and flexor *digitorum profundus* (FDP) of the second and at times part of the third finger [33]. The PQ is also affected but as the PTM remains intact the pronation is preserved. Patients notice lack of muscle power in the pinch between the thumb and index finger [62]. This induces difficulties on the writing hand [62], and the patient cannot make the OK sign [58]. There is no sensory deficit [61]. The causes are a tendinous origin of the PTM deep head or the third finger FDS, collateral ulnar vessel thrombosis, an accessory head of the FPL (Gantzer's muscle), aberrant radial arteries or an enlarged bicipital bursa encroaching on the MN near the AIN site of origin [63]. Electrodiagnostic studies can confirm the diagnosis. If conservative treatment fails, surgical decompression is indicated [64]. The results are usually satisfactory provided that the nerve is freed on time.

2.4. Ulnar nerve compression at the elbow

Cubital tunnel syndrome (CubTS) is the entrapment of the UN at the elbow [65]. It is the most common site of UN entrapment and the second most common in the upper extremity nerve

[66]. Its estimated incidence is 25 new cases/100,000 inhabitants/year [67, 68], affecting males more often than females [68–72]. It is more common in jobs with constant leaning on the elbow (i.e. book keepers, drivers resting the elbow on the window frame) [73], gripping tools (gardeners, farmers, builders) [74], professional motorbike runners, cyclist [75, 76], repetitive elbow flexoextension [73, 74] and in floor cleaners [73, 77]. It is also more frequent in some systemic disorders like diabetes mellitus [16], acromegaly [78], rheumatoid arthritis [79] or amyloidosis [80]. CTS and CubTS in the same arm is not a rare finding [81–83].

Its clinical presentation consists of pain, sensory loss, paraesthesias, motor weakness and muscle atrophy at the forearm ulnar side and fourth and fifth fingers [33]. If untreated, it can lead to lack of sensation and muscle power, as well as pain and clumsiness in the affected hand [70]. Patients often complain of a dull pain at the elbow with shock-like sensations with any mild pressure or blow on this area. Some patients notice no sensory symptoms because of progressive weakness in the fourth and fifth fingers accompanied by muscle atrophy of the hand intrinsic muscles (**Figures 2E** and **F**) [84]. Symptoms get worse with activity and on flexing the elbow.

On clinical examination, the fifth finger remains in abduction due to weakness of the fourth palmar interosseous muscle (*Wartenberg sign*) (**Figure 2A**). This finger stays behind and out when the patient is attempting to put his/her hand inside the pocket [33]. The *Froment sign* is the flexion of the distal phalanx of the thumb when attempting to hold a piece of paper (**Figure 2C**). It is due to weakness of adductor *pollicis*, flexor *pollicis brevis* and first dorsal interosseous muscle, being substituted by the action of the FPL [33]. The weakness of the *interossei* and lumbrical muscles induces metacarpophalangeal joint hyperextension with flexion of the interphalangeal joint of the fourth and fifth fingers, creating the *'claw hand', 'main en griffe' or Duchenne sign* (**Figure 2B**) [33]. Contrariwise to the hand of benediction seen with medial nerve damage at the forearm, the ulnar claw hand is due to the impossibility of the fourth and fifth fingers to extend. Meanwhile in the hand of benediction it is impossible to flex the thumb, index and middle digits when attempting to make a fist. In CubTS the weakness and atrophy of the first dorsal interosseous muscle (**Figure 2D**) is much more severe and earlier than the weakness and atrophy of the abductor *digiti minimi* (ADM) [33].

The most common site of UN entrapment is the retroepicondylar groove followed by the cubital tunnel 1.5–3 cm distal to the epicondyle [3]. About 40% of the cases are idiopathic [33]. The causes of compression are a bulky triceps muscle [85], the *anconeus epitrochlearis* muscle [86], fibrous bands bridging between the medial epicondyle and the olecranon [87], Osborne's fascia [88] or the point where the UN crosses under the two heads of the flexor *carpi ulnaris* muscle [89]. It can occur after trauma or a protracted wrong position of the arm. It is the most common entrapment syndrome after anaesthesia for surgical procedures, particularly if they are long [90].

The diagnosis is based on the symptoms. Electrodiagnostic studies confirm the diagnosis and rule out other medical conditions (i.e. C_8 radiculopathy) [3].

Some patients may improve with conservative measures like avoiding external elbow pressure, using a night time split to keep the elbow extended or stopping any occupational activity that might be causing the disease. If that is not enough or the patient presents with muscle weakness and atrophy, a surgical decompression is indicated.

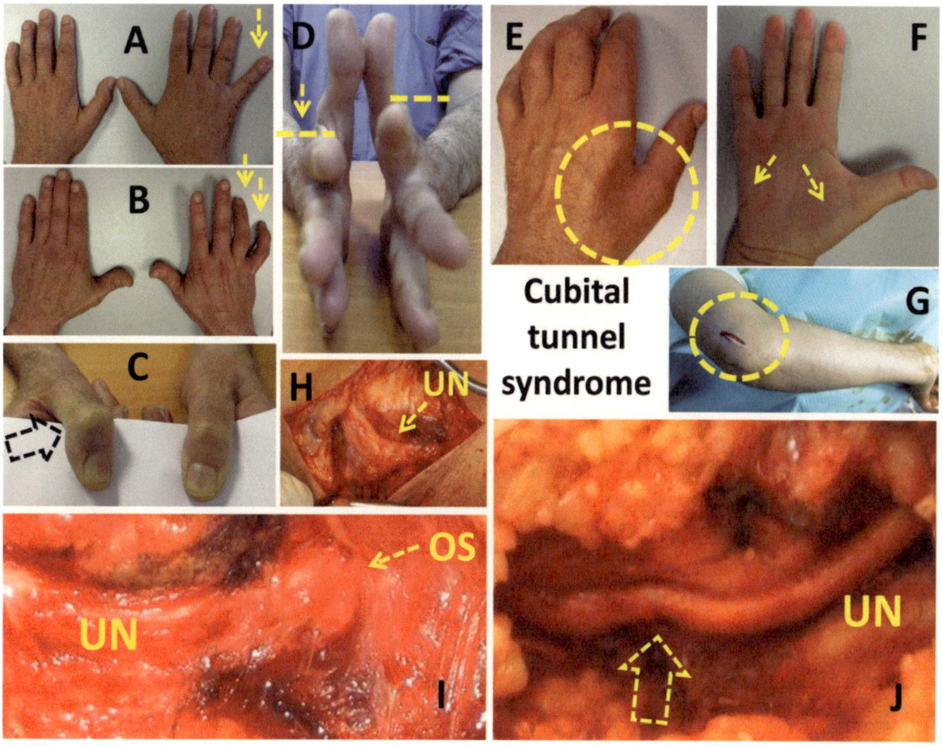

Figure 2. Cubital tunnel syndrome. (A) Wartenberg sign; (B) claw hand; (C) Froment sign; (D) weakness of the first interosseous muscle; (E) atrophy thenar eminence; (F) atrophy thenar and hypothenar eminences; (G) surgical incision; (H) ulnar nerve (UN) compression by the (I) Osborne's arcade (OS); (J) ulnar nerve fully decompressed with a compression mark at the retroepicondylar tunnel (depicted by the arrow).

The techniques for UN decompression at the elbow are medial epicondylectomy, *in situ* decompression and transposition. The epicondylectomy is not popular anymore. Several clinical comparative studies [91–94], meta-analyses [95–97] and prospective randomized trials [98, 99] have shown that *in situ* CubTS decompression (**Figures 3A–D**) is just as effective as transpositions (**Figures 4A–D**) provided there is no UN subluxation on elbow flexoextension [97, 99]. Advantages of *in situ* decompression are smaller surgical incisions, less risk of medial antebrachial cutaneous nerve damage [100], no UN devascularisation [97], shorter operating time [101], smaller costs [99] and a faster recovery [102]. Transpositions need more surgical time, are more expensive and have more complications than *in situ* decompression [99], but can be used in the case of failure [102, 103]. It is imperative to avoid damaging the medial antebrachial cutaneous nerve or its branches regardless of the technique, as injury will induce post-operative neuropathic pain in the elbow area [104–106]. The *in situ* decompression can be performed open field or endoscopic. The latter shows a higher rate of post-operative surgical field haematoma [107]. The clinical results are equivalent for both procedures [100, 107, 108]. In recurrent cases, the most frequent finding is incomplete decompression either at the distal flexor-pronator muscle group or proximally at the intermuscular septum [109, 110].

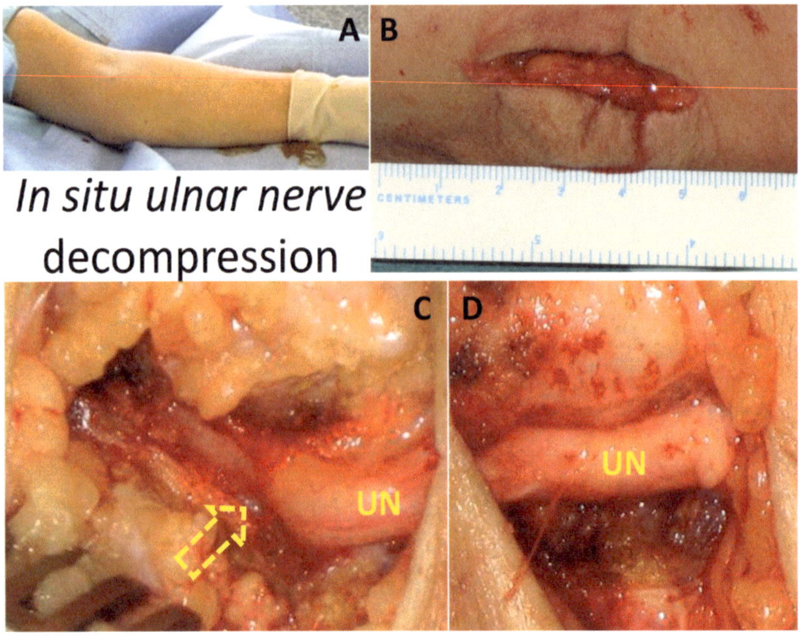

Figure 3. *In situ*, ulnar nerve decompression. (A) patient position; (B) skin incision; (C) ulnar nerve (UN) decompression, the arrow points to the entrapment point at the retroepicondylar tunnel; (D) ulnar nerve fully decompressed.

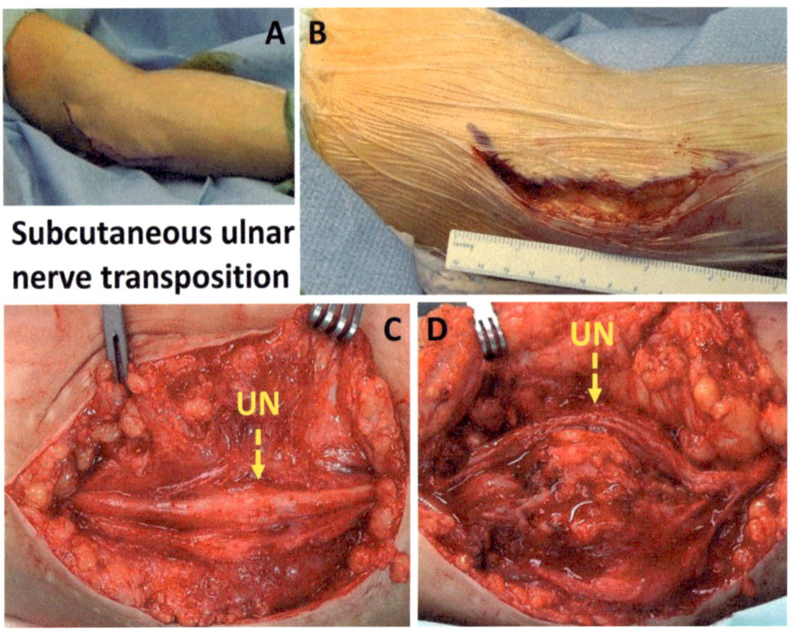

Figure 4. Cubital tunnel syndrome. Subcutaneous ulnar nerve transposition. (A) patient position; (B) skin incision; (C) ulnar nerve (UN) decompressed; (D) ulnar nerve transposed subcutaneously.

Pre-operative and intra-operative electrophysiological inching studies have found that the compression point is at or immediately proximal to the cubital tunnel [87, 111–113], less often at the Osborne's arcade but not proximally at the intermuscular septum [82]. Others with endoscopic assistance have reported no nerve constriction beyond 4 cm distally or proximally to the retroepicondylar tunnel [100]. So, extensive proximal decompression seems futile [100, 114]. Unsatisfactory results have been related to concomitant undiagnosed CTS or to weight gain [107].

2.5. Ulnar nerve compression at the hand

It is an uncommon site for UN entrapment (**Figure 5A**). Depending on the exact point of compression, it can be classified into five types [88, 115, 116]. In type I, the compression is proximal to Guyon's canal with involvement of the superficial sensory, hypothenar motor, as well as deep motor branch. In type II, the compression is inside the canal and only the superficial sensory branch is affected. In type III, the compression is distal to the sensory branch with involvement of the hypothenar and deep motor branch proximal to the branch for the ADM. In type IV, the compression is distal to the superficial sensory and the hypothenar branch, so only the deep motor branch is affected. In type V, there is compression to the deep motor branch just proximal to the adductor *pollicis* and first dorsal interosseous muscles.

Usually there is the antecedent of an acute trauma [58] or chronic compression (cyclists) [76]. In other cases, there is a structural lesion in the area compressing the nerve, most commonly a ganglion cyst [116]. In cases of repetitive compression (i.e. cyclists), removal of the offending activity can be tried. If there is a lesion it has to be removed before irreversible UN damage develops (**Figures 5B–E**) [116].

2.6. Radial nerve (RN) entrapment syndromes

Its entrapment points are [3] at the spiral groove by the intermuscular septum between the triceps and brachialis (BaM) muscles, at the proximal forearm by the ligament of Frohse (posterior *interosseous* nerve or PIN), between the two heads of the *supinator* muscle (PIN) and by edge of the *brachioradialis* muscle (BRM) in the distal forearm (superficial cutaneous branch of the RN). It is the third most common upper limb entrapment [3].

The *Saturday Night palsy or Honeymoon palsy* is the most common compression of the RA occurring at the spiral groove [117]. It is usually due to local pressure on the posterior aspect of the arm under anaesthesia, drug intoxication (i.e. alcohol) or profound sleep with the arm over a hard surface or under the body of somebody else (Honeymoon palsy) [118]. In most cases, it improves spontaneously [117].

The *radial tunnel syndrome* is the entrapment of the deep branch of the RA [119]. This tunnel begins where the deep branch of the RN crosses over the radiohumeral joint, ending where this nerve becomes the PIN below the supinator muscle (SM) [33]. It has an average length of 5 cm. Its lateral wall is formed by the muscles BRM, extensor *carpi radialis brevis* (ECRB), BaM and extensor *carpi radialis longus* (ECRL). Its medial wall is created by the biceps tendon and the *BaM*. The floor of this tunnel is created by the radiocapitellar joint [120]. The compression can be due to the arcade of Frohse (tendinous border of the superficial layer of the SM),

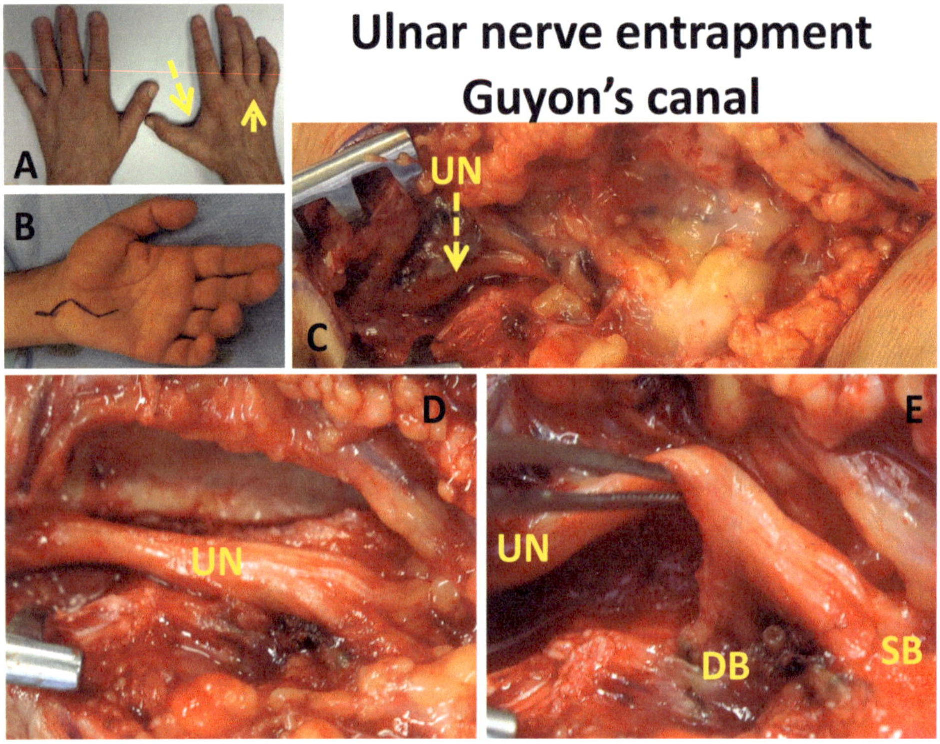

Figure 5. Ulnar nerve entrapment at wrist. (A) Atrophy thenar eminence and claw hand; (B) skin incision; (C) ulnar nerve (UN) exposed in the distal third of the forearm; (D) ulnar nerve exposed in the wrist; (E) ulnar nerve deep (DB) and superficial branch (SB) liberated.

the superomedial border of the ECRB, the inferior border of the superficial layer of the SM, fibrous bands at the radio-humeral joint and some radial vessels with a recurrent direction and a fibrous septum between the BRM and BaM muscles [119, 121, 122]. The treatment is conservative [123] avoiding the movements that induce pain but if it persists surgical exploration may be justified [124].

The *PIN entrapment* induces pain in the lateral aspect of the arm and forearm [125] and weakness in extension in all fingers (thumb included) and in thumb abduction (**Figure 6A**). There is no wrist drop because the ECRB is spared but on wrist extension the extensor *carpi ulnaris* weakness induces radial deviation [33]. There is no sensory deficit as the superficial RN is not involved. Symptoms exacerbate on hand or forearm repetitive movements [126]. Symptoms can be reproduced by direct pressure on the radial aspect of the forearm at 6 cm distal to the epicondyle. The provocative test consists in forceful hand supination with the shoulder in adduction and the elbow at 90° flexion or with extension of the middle finger [58]. It can be due to entrapment by the arcade of Frohse or on its way between both heads of the SM [126] (**Figure 6C**), ganglion cysts [125] or benign tumours (lipoma the most frequent) [127]. When symptomatic its treatment is surgical decompression [128] (**Figures 6B–E**).

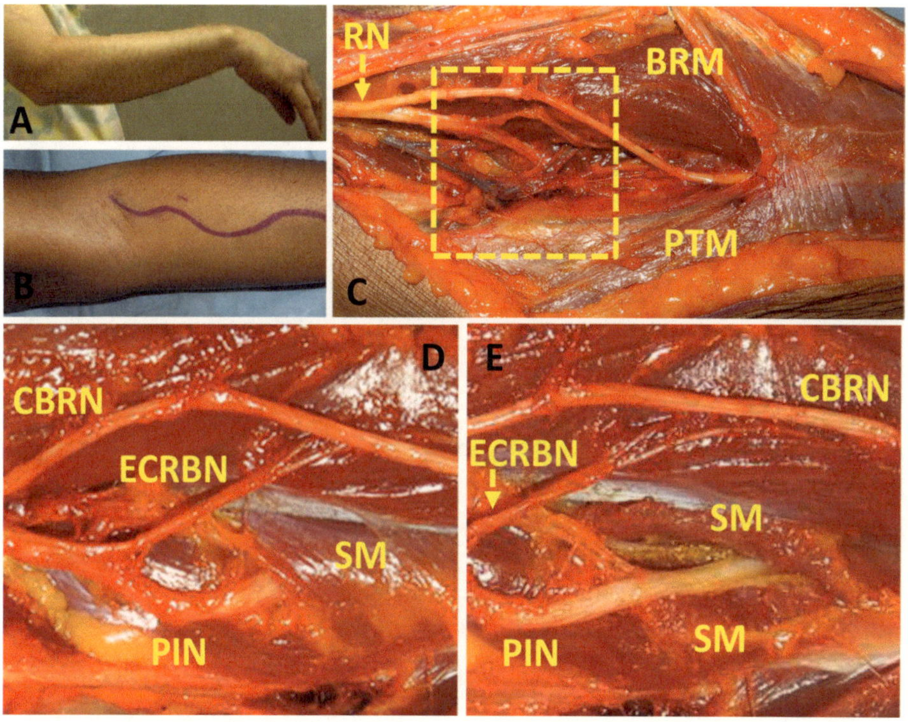

Figure 6. Posterior interosseous nerve entrapment. (A) paralysis of finger extension; (B) skin incision; (C) the radial nerve (RN) is found between the *Brachioradialis* (BRM) and Pronator *Teres* (PTM) muscles; (D) the posterior interosseous nerve (PIN) is compressed between both heads of the *Supinator* muscle (SM), cutaneous branch radial nerve (CBRN), extensor carpi radialis brevis nerve (ECRBN); (E) the supinator muscle has been sectioned, freeing the posterior interosseous nerve.

Cheiralgia paraesthetica or Wartenberg syndrome is the entrapment of the superficial cutaneous branch of the RN. There is numbness and pain in the dorsum of the hand, the thumb, index and middle fingers, accompanied by dysaesthesia, burning sensation and hyperesthesia [58]. The entrapment points are at the forearm between the tendons of the BRM and ECRB or at the wrist at its exit from beneath the fascia to the subcutaneous layer in the site where the fascia joins the tendons of the BRM and ECRL [129]. In the first case, it can be due to repetitive pronation and supination [58]. In this second case, it is usually due to external compression by handcuffs [130] or a wristwatch [131]. The best provocative test is to ask the patient to place their arm under maximum pronation with the wrist flexed to the ulnar side [132]. The Finkelstein test is positive, inducing confusion with a De Quervain syndrome [58]. The treatment is the removal of the cause. In case of persistence, surgical decompression of this nerve, particularly as it crosses the edge of the tendon of the BRM muscle, is indicated [131, 133] (**Figures 7A–D**).

2.7. Suprascapular nerve entrapment

It can be trapped at the suprascapular and spinoglenoid notches where the nerve is fixed by ligaments in a bony canal [134]. The first symptom is pain localized in the posterior aspect of the shoulder that gets worse with activity, when lying on the affected area, or by

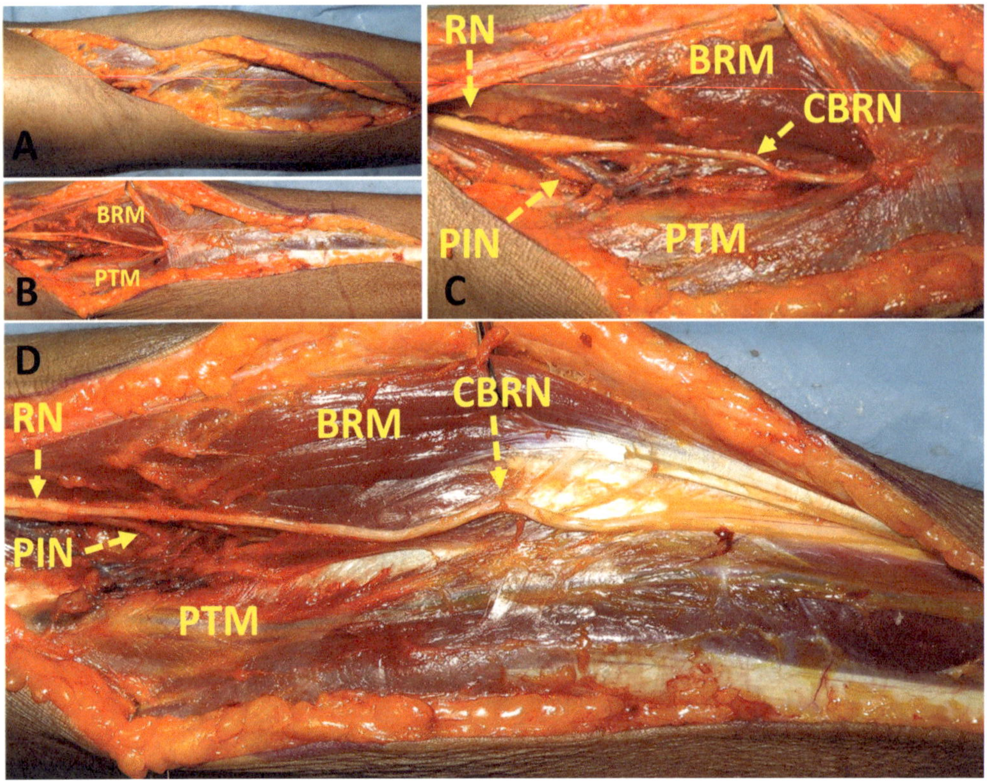

Figure 7. Superficial cutaneous nerve branch radial nerve entrapment. (A) skin incision; (B) approach between the brachioradialis (BRM) and pronator teres (PTM) muscles; (C) and (D) exposure and liberation of the superficial cutaneous radial nerve branch (CRNB), radial nerve (RN), posterior interosseous nerve (PIN).

shoulder adduction crossing the midline with the extended arm [33]. The weakness and atrophy of the supra and infraspinatus muscles induce paresis of shoulder abduction and external rotation.

On clinical examination, the affected shoulder is lower than the healthy one and the scapular muscles are atrophied (**Figure 8A**). The patient has difficulty raising the outstretched arm above the horizontal (**Figures 8B** and **C**).

Its treatment is surgical with section of the ligament that closes the suprascapular notch at the superior aspect of the scapula. It can be done open field [135] (**Figures 8D** and **E**) or endoscopically [136] with similar outcomes.

2.8. Thoracic outlet syndrome

There is pain in the inner aspect of the arm and forearm, sometimes reaching the fourth and fifth fingers [137]. This pain gets worse when lifting the arm above the horizontal [138]. Sometimes there is associated hand muscle atrophy [139]. Claw hand deformity can be

http://dx.doi.org/10.5772/67946

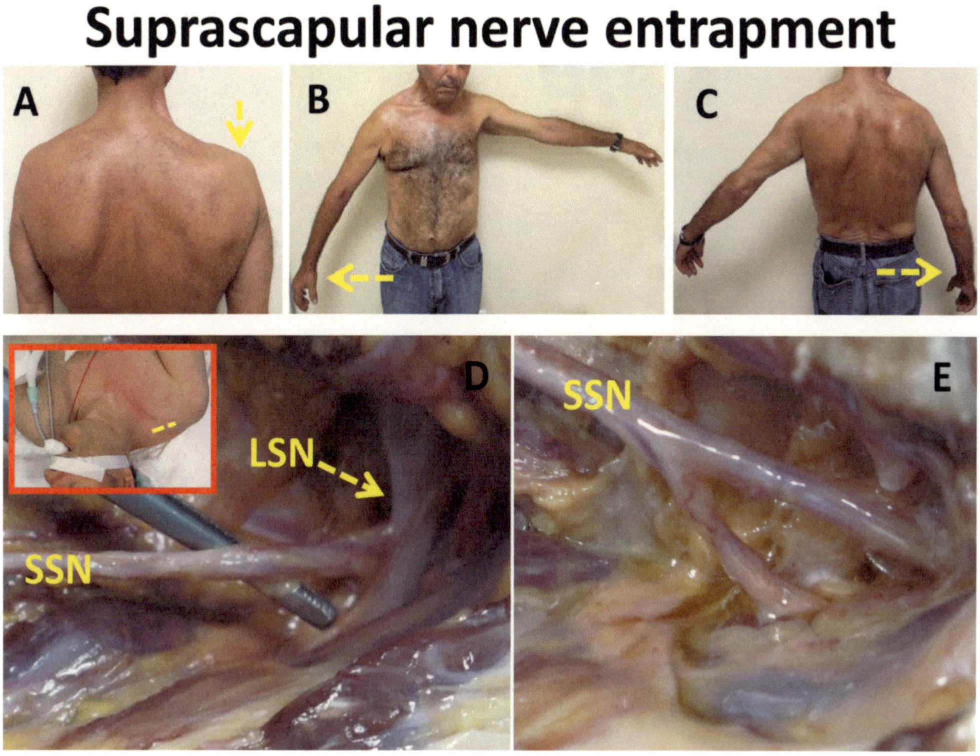

Figure 8. Suprascapular nerve entrapment. (A) Atrophy supra and infraspinatus muscles; (B) and (C) weakness of shoulder abduction; (D) suprascapular nerve (SSN) exposed with the ligament for the suprascapular notch (LSN) intact; (E) suprascapular nerve after removal of the ligament for the suprascapular notch.

present in the long protracted cases [138]. The neurogenic type has an incidence of one case per million inhabitants [137]. It can be due to hypertrophy of some muscles at the root of the arm (typical of ceiling painters or swimmers) or to the existence of a cervical rib or fibrous ligament at the same point [137].

The clinical presentation is pins and needles with overhead activities (like painting a ceiling) [140], carrying heavy objects with the arms hanging down [138], combing hair and applying makeup [4].

In the *Roos' elevated arm stress test*, the shoulder is abducted 90° and kept in external rotation with the elbow in 30° flexion. This provocative test is positive if opening and closing the hand for 1 min reproduces the symptoms [141].

In case of poor response to conservative treatment, surgical decompression is in order. The two possibilities are the transaxillary removal of the first rib [142] or supraclavicular scalenectomy [143] (**Figures 9A–D**). This depends on the causative mechanism and the surgeon's preferences but the supraclavicular approach offers a better chance of solving any causative abnormality [143].

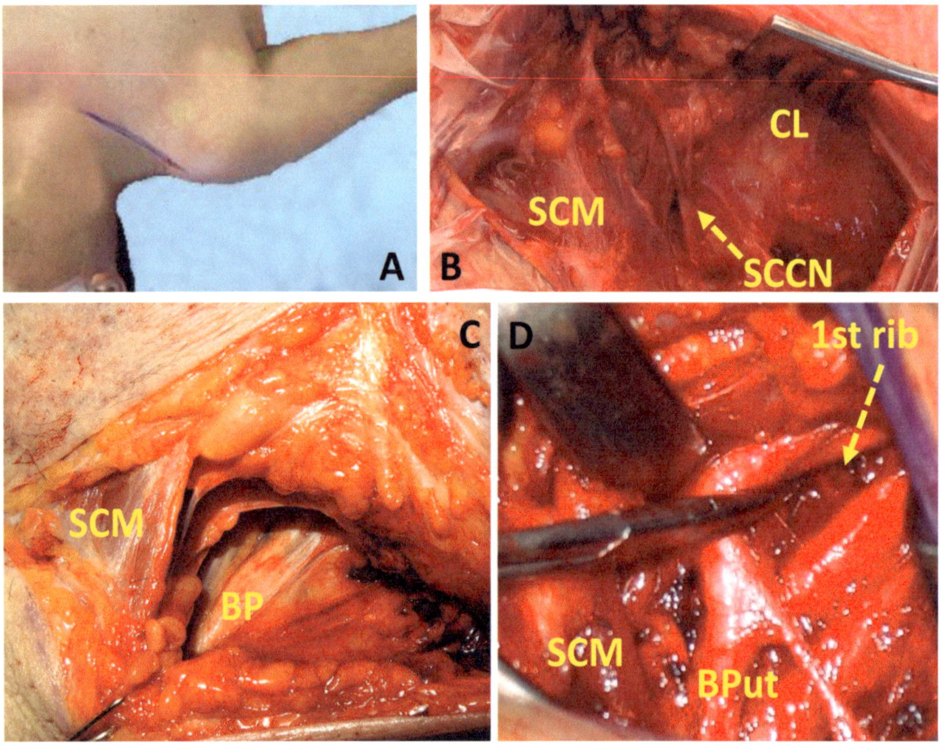

Figure 9. Thoracic outlet syndrome. (A) skin incision; (B) supraclavicular cutaneous nerve (SCCN), sternocleidomastoid muscle (SCM), clavicle (CL); (C) brachial plexus (BP) exposed; (D) first rib exposed, brachial plexus upper trunk (BPut).

3. Lower limb entrapment syndromes

3.1. Meralgia paraesthetica

The femoral cutaneous nerve is a purely sensory nerve which runs usually medial to the ASIS. The name meralgia paraesthetica comes from the Greek, *meros* meaning thigh and *algos* meaning pain. The clinical presentation is pain, paraesthesia, numbness and hypersensitivity in the anterolateral side of the thigh down to the knee (**Figures 10A** and **B**) [144]. There are no motor signs and if present a different medical condition should be suspected [145]. The pain worsens when standing or walking and improves on sitting [146]: it is aggravated on leg extension and improved on knee flexion [146]. Its estimated incidence is 36.2 cases/100,000 habitants/year in the USA [147] and 43 cases/100,000 persons/year in Europe [145]. Meralgia paraesthetica is more frequent in obese people, in the fourth–six decades of life [145, 147, 148] and in diabetics (seven times more than in the general population) [147, 148]. Although many cases are idiopathic oftentimes its cause is an external compression of the nerve as it passes underneath or through the anterior inguinal ligament at its origin on the ASIS [149]. This can be due to either internal causes like a bulging abdomen [150], obesity [147], pregnancy [151], ascites [152], external agents like tight clothes [153] or belts resting on the outer aspect of the thigh repeatedly while standing (typical of hairdressers) or due to a prolonged position (lithotomy posture, cycling) [144]. It can also be induced by a pelvic fracture with psoas haematoma or by surgical procedures such as hip [154] or knee replacements [155] or an aortofemoral bypass [144, 147]. It has also been related to acute seat belt compression in car accidents [144].

The clinical presentation in the absence of motor signs helps to make the diagnosis. Electrodiagnostic studies can rule out confounding conditions [144, 156], but ultrasonography is very useful, particularly in obese patients [156, 157].

Initially, the treatment is to remove the compressing agent and/or lose weight. If insufficient, the area can be infiltrated with a local anaesthetic agent and corticosteroids [158]. The rebel cases require surgical treatment with nerve decompression (**Figures 10C–E**) or neurectomy [159].

3.2. Peroneal nerve entrapment

It is the most common lower limb entrapment neuropathy [162]. It is a mixed nerve that runs at the fibular head, reaching the anterior compartment of the leg distal to the knee [160]. At that level, it lies between the skin and bone. This makes it very sensitive to trauma or pressure,

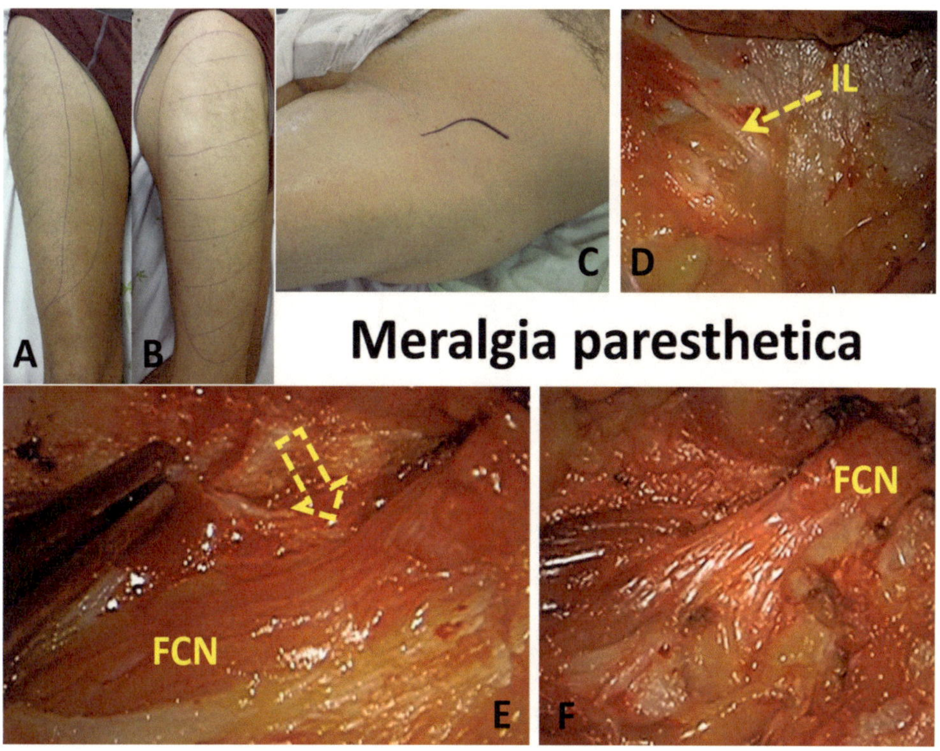

Figure 10. Femoral cutaneous nerve entrapment (meralgia paraesthetica). (A) and (B) area of sensory disturbance; (C) skin incision; (D) subcutaneous tissues with inguinal ligament (IL); (E) femoral cutaneous nerve (FCN) with the arrow pointing at the entrapment point under the inguinal ligament; (F) femoral cutaneous nerve fully decompressed.

particularly in bedridden lean patients [161]. It can also be due to mass lesions (i.e. ganglion cyst of the tibiofibular joint) or associated with systemic diseases (diabetes and vasculitis) [162]. It is more frequent in occupations requiring people to squat for long periods of time (strawberry pickers, farm workers and carpet layers) [163] or that sit crossing their legs [164].

The clinical presentation is pain at the fibular head and loss of strength in dorsiflexion (**Figure 11A**), which causes the foot to drag when walking. The patient notices foot slap with steppage gait and wearing the tip of the shoe as well as a sensory loss on the dorsal aspect of the foot between the first and second toe [165].

The treatment is surgical decompression (**Figures 11B–G**) [161, 166].

3.3. Anterior tarsal tunnel

It is the entrapment of the deep peroneal nerve. The clinical presentation is pain in the dorsum of the foot associated with sensory loss in the first foot web space [167]. The treatment is initially conservative, but surgical decompression with extensive fascial opening might be needed [168].

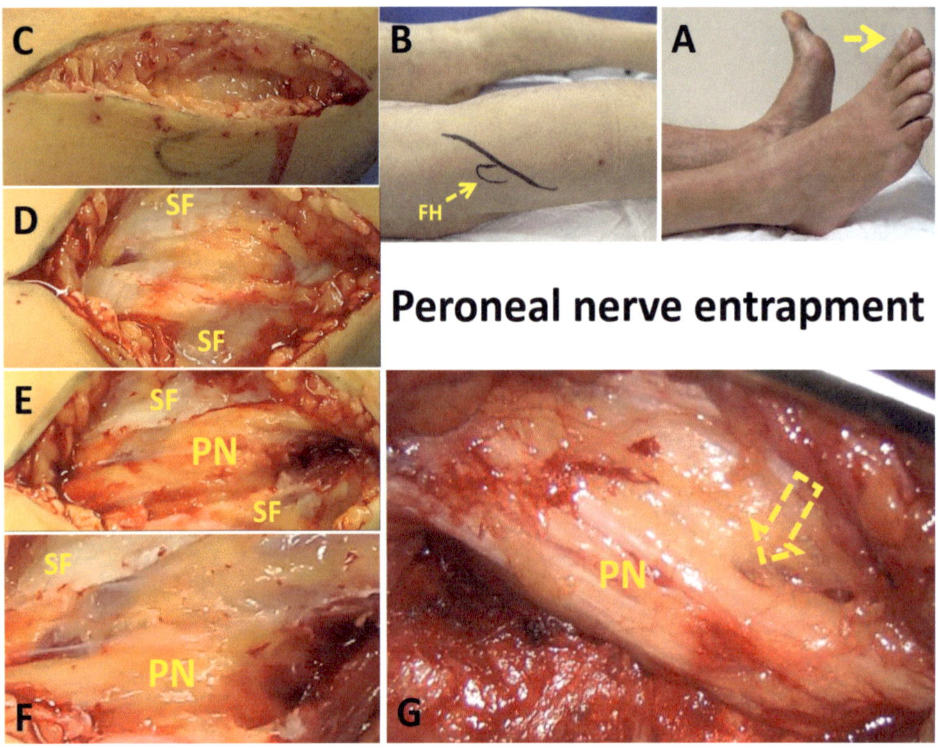

Figure 11. Peroneal nerve entrapment at the level of the fibular head. (A) foot dorsiflexion weakness; (B) skin incision and fibular head (FH) depicted; (C)–(G) steps in peroneal nerve (PN) decompression, subcutaneous fascia (SB). The arrow points at the marked impinged on the nerve by the compressing fascial band.

3.4. Tarsal tunnel syndrome

It is due to compression of the posterior tibial nerve at the tarsal tunnel behind the foot medial malleolus. It is very uncommon. Many cases are idiopathic (20–46%) [169]. Contributing factors are ankle sprain and fracture, tight-fitting foot wear and space occupying lesions [170]. The clinical presentation is pain, paraesthesia and numbness in the sole of the foot. This symptoms get worse on standing, walking and at night time [2]. The sensory loss affects the sole of the foot sparing the heel, supplied by the calcaneal branch [171]. The diagnosis is identified with the clinical presentation. Electrodiagnostic studies can be useful to rule out confounding medical conditions [172]. Ultrasonography [173] and MRI [174] can rule out associated space occupying lesions.

Its initial treatment is rest and anti-inflammatories, but if there is no improvement or relapse after an initial response, surgical decompression may be necessary. This can be done endoscopically [175], but for a good decompression, especially in the distal part, an open approach gives better results [176] (**Figures 12A–D**).

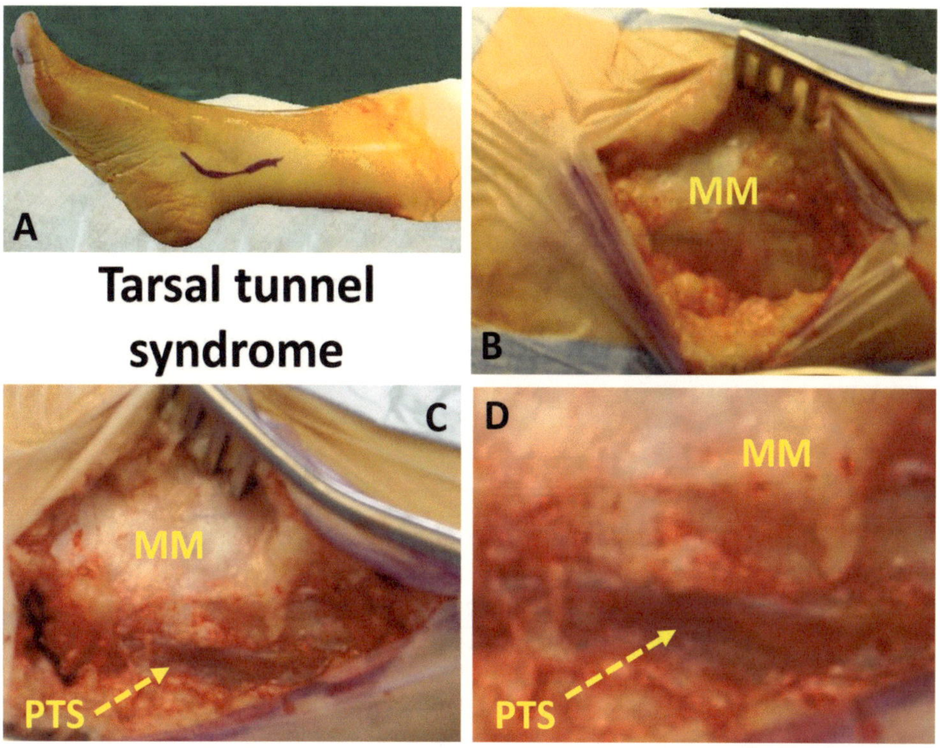

Figure 12. Tarsal tunnel syndrome. (A) skin incision; (B) exposure medial malleolus (MM); (C) section of roof tarsal tunnel, posterior tibial structures (PTS); (D) exposure posterior tibial structures.

3.5. Piriformis syndrome

It is a very rare disorder in which *the sciatic nerve is compressed by the piriformis muscle*. This is a flat, pyramid-shaped *muscle located deep in the gluteal region*, running between the femur and the iliac bone. It helps in hip external rotation [177]. Common sciatic nerve pain (i.e. lumbar disc hernia) is much more prevalent than this syndrome.

The conditions associated with this syndrome are sitting for extended periods of time, sitting with a large wallet in the rear pocket, repeated forward movements, running, bicycling, stiff sacroiliac joints, foot overpronation, Morton's toe (the second toe is longer than the first one) and after a fall on the buttocks [177–179]. Approximately 50% of the cases are caused by trauma and the rest are spontaneous [178].

The symptoms are sciatica-like pain. Pain starts in the gluteal area and may travel through the back of the thigh and calf up to sole the foot. Patients might experience tingling, numbness, burning sensation and weakness. The sciatic pain aggravates with sitting or with activities that press the piriformis against the sciatica nerve, such as running, cycling or hose riding [178].

The diagnosis is usually made through physical examination. Certain tests may elicit sciatica nerve pain indicating the presence of the syndrome, especially internal rotation of the hip with the knee in full extension.

On MRI examination, it is possible to see the sciatic nerve with oedema when crossing under the piriformis muscle.

Conservative treatment is initially recommended. Alternate ice and heat treatment may provide relief. Ultrasound penetrates deep into the muscle alleviating the sciatica nerve pain. Stretching exercises to target the piriformis, hamstrings and hip muscles, will help increase the range of motion and decrease the sciatic nerve pain.

If all these treatments prove unsuccessful, injection of botulinum toxin in the piriformis muscle [180] under CT or MRI guidance can be attempted. In the case of failure, surgical decompression removing the piriformis muscle or the offending fibrous band could be indicated [179, 181] (**Figures 13A–E**). The results are inconsistent.

3.6. Pudendal nerve entrapment

It induces pain in the genital and sometimes gluteal areas [182]. The pain worsens with local pressure, sitting, defecating, and urinating and with sexual intercourse [183]. It is constant, intense and burning. The cause can be local pressure induced by repeated cycling [184] or by horse riding. The problem is that most patients are diagnosed late. Once suspected, it can be confirmed with electrodiagnostic studies [185]. When conservative treatments [186] fail, surgical decompression should be considered [187] (**Figures 14A–E**). The results are often poor, at times because patients are diagnosed much too late due to lack of awareness in the medical world.

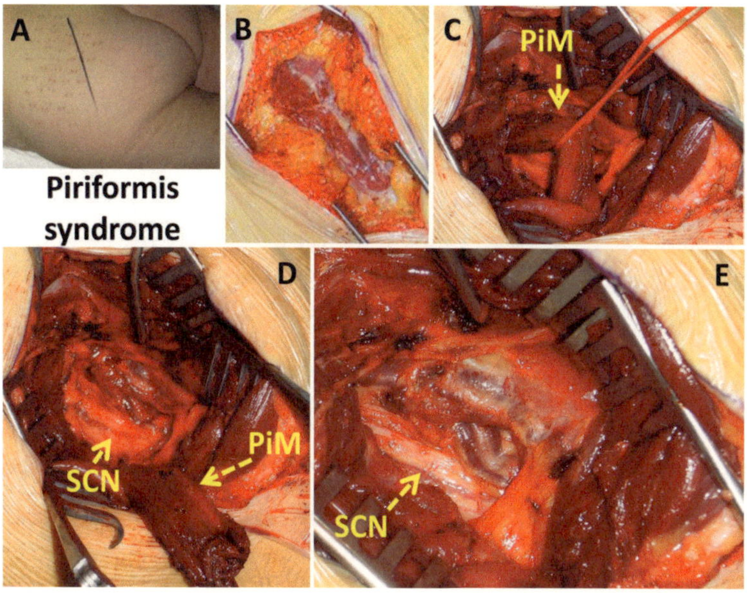

Figure 13. Piriformis syndrome with sciatic nerve (SCN) entrapment. (A) Skin incision; (B) gluteus *maximus* muscle exposed; (C) piriformis muscle (PiM) isolated; (D) piriformis muscle sectioned exposing the sciatic nerve (SCN); (E) sciatic nerve free.

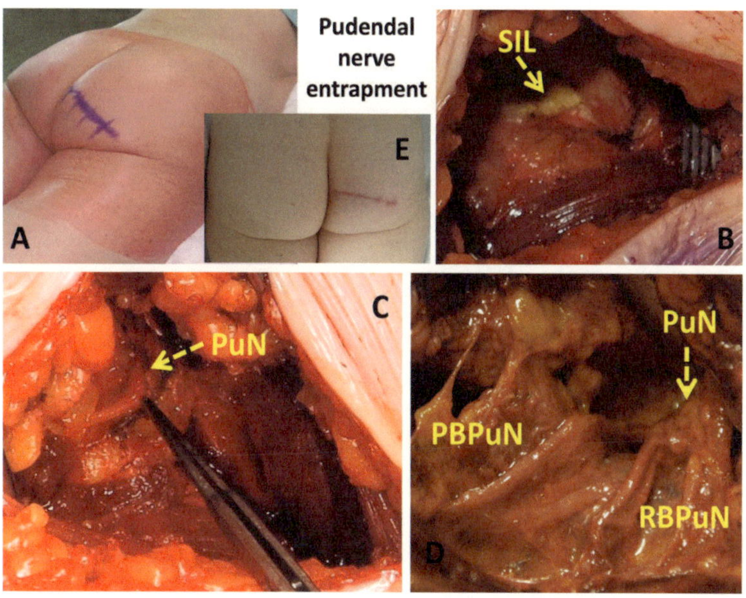

Figure 14. Pudendal nerve entrapment. (A) skin incision; (B) gluteus maximus muscle dissection exposing the sacroischias ligament (SIL); (C) the pudendal nerve (PuN) is exposed after sectioning the sacroischial ligament; (D) the pudendal and its branches are freed (RBPuB, rectal branch pudendal nerve; PBPuN, perineal branch pudendal nerve); (E) post-operative scar.

4. Conclusion

Nerve entrapments syndromes are more frequent than currently thought. Their awareness is essential to diagnose and treat the patient on time. Although almost any nerve can suffer an entrapment syndrome, some are more common than others. The most frequent is CTS, followed by meralgia paraesthetica and ulnar nerve entrapment at the elbow. The clinical presentation is pain, paraesthesia and muscle power loss in the distribution of the affected nerve. Many cases are idiopathic, but others are induced by internal or external compressing mechanisms. Some systemic conditions are associated with an increased incidence of these syndromes. The clinical presentation together with the electrodiagnostic studies help in the diagnosis. The ultrasonography and the MRI are also helpful but not used so regularly.

In many cases, conservative medical treatment is sufficient. When it is not, surgical decompression has to be performed.

Open and endoscopic approaches are available. In each case, we will have to see which shows better outcomes. A few cases with persistent pain after surgical decompression might benefit from peripheral nerve neurostimulation.

Acknowledgements

We thank the Department of Human Anatomy and Embryology of the Faculty of Medicine of the University of Valencia, particularly to the laboratory curators Lucia and Carmina and to Dr. Tomás Hernández Gil de Tejada, and to all personnel of the *Instituto de Medicina Legal de Valencia*, for their assistance in this study.

Appendices and nomenclatures

AIN	Anterior *interosseous* syndrome
ASIS	Anterior superior iliac spine
BaM	*Brachialis* muscle
BRM	*Brachioradialis* muscle
CubTS	Cubital tunnel syndrome
CTS	Carpal tunnel syndrome
ECRB	Extensor *carpi radialis brevis*
ECRL	Extensor *carpi radialis longus*
FDS	Flexor *digitorum superficialis*
FDP	Flexor *digitorum profundus*
FPL	Flexor *pollicis longus*

http://dx.doi.org/10.5772/67946

MN	Median nerve
PIN	Posterior interosseous nerve (PIN)
PQ	Pronator *quadratus* muscle
PTM	Pronator *teres* muscle
RN	Radial nerve
SM	*Supinator* muscle
UN	Ulnar nerve

Author details

Vicente Vanaclocha-Vanaclocha[1*], Nieves Sáiz-Sapena[2], Jose María Ortiz-Criado[3] and Nieves Vanaclocha[4]

*Address all correspondence to: vvanaclo@hotmail.com

1 Department of Neurosurgery, Hospital General Universitario, Valencia, Spain

2 Hospital 9 de Octubre, Valencia, Spain

3 Universidad Católica San Vicente Mártir, Valencia, Spain

4 Hospital General Universitario Valencia, Spain

References

[1] Papanicolaou GD, McCabe SJ, Firrell J. The prevalence and characteristics of nerve compression symptoms in the general population. J Hand Surg. 2001 May;26(3):460–6.

[2] McSweeney SC, Cichero M. Tarsal tunnel syndrome—a narrative literature review. Foot (Edinb Scotland). 2015 Dec;25(4):244–50.

[3] Arnold WD, Elsheikh BH. Entrapment neuropathies. Neurol Clin. 2013 May;31(2):405–24.

[4] Tang DT, Barbour JR, Davidge KM, Yee A, Mackinnon SE. Nerve entrapment: update. Plast Reconstr Surg. 2015 Jan;135(1):199e–215e.

[5] Ettema AM, Amadio PC, Zhao C, Wold LE, O'Byrne MM, Moran SL, An K-N. Changes in the functional structure of the tenosynovium in idiopathic carpal tunnel syndrome: a scanning electron microscope study. Plast Reconstr Surg. 2006 Nov;118(6):1413–22.

[6] Latko WA, Armstrong TJ, Franzblau A, Ulin SS, Werner RA, Albers JW. Cross-sectional study of the relationship between repetitive work and the prevalence of upper limb musculoskeletal disorders. Am J Ind Med. 1999 Aug;36(2):248–59.

[7] Wieslander G, Norbäck D, Göthe CJ, Juhlin L. Carpal tunnel syndrome (CTS) and exposure to vibration, repetitive wrist movements, and heavy manual work: a case-referent study. Br J Ind Med. 1989 Jan;46(1):43–7.

[8] Fox IK, Mackinnon SE. Adult peripheral nerve disorders: nerve entrapment, repair, transfer, and brachial plexus disorders. Plast Reconstr Surg. 2011 May;127(5):105e–118e.

[9] Mills KR. The basics of electromyography. J Neurol Neurosurg Psychiatry. 2005 Jun;76 Suppl 2:ii32–35.

[10] Rempel DM, Diao E. Entrapment neuropathies: pathophysiology and pathogenesis. J Electromyogr Kinesiol Off J Int Soc Electrophysiol Kinesiol. 2004 Feb;14(1):71–5.

[11] Kim S, Choi J-Y, Huh Y-M, Song H-T, Lee S-A, Kim SM, Suh J-S. Role of magnetic resonance imaging in entrapment and compressive neuropathy—what, where, and how to see the peripheral nerves on the musculoskeletal magnetic resonance image: part 1. Overview and lower extremity. Eur Radiol. 2007 Jan;17(1):139–49.

[12] Kim S, Choi J-Y, Huh Y-M, Song H-T, Lee S-A, Kim SM, Suh J-S. Role of magnetic resonance imaging in entrapment and compressive neuropathy—what, where, and how to see the peripheral nerves on the musculoskeletal magnetic resonance image: part 2. Upper extremity. Eur Radiol. 2007 Feb;17(2):509–22.

[13] Koenig RW, Pedro MT, Heinen CPG, Schmidt T, Richter H-P, Antoniadis G, Kretschmer T. High-resolution ultrasonography in evaluating peripheral nerve entrapment and trauma. Neurosurg Focus. 2009 Feb;26(2):E13.

[14] Mackinnon SE. Double and multiple "crush" syndromes. Double and multiple entrapment neuropathies. Hand Clin. 1992 May;8(2):369–90.

[15] Upton AR, McComas AJ. The double crush in nerve entrapment syndromes. Lancet Lond Engl. 1973 Aug 18;2(7825):359–62.

[16] Rota E, Morelli N. Entrapment neuropathies in diabetes mellitus. World J Diabetes. 2016 Sep 15;7(17):342–53.

[17] Atroshi I, Gummesson C, Johnsson R, Ornstein E, Ranstam J, Rosén I. Prevalence of carpal tunnel syndrome in a general population. JAMA. 1999 Jul 14;282(2):153–8.

[18] de Krom MC, Knipschild PG, Kester AD, Thijs CT, Boekkooi PF, Spaans F. Carpal tunnel syndrome: prevalence in the general population. J Clin Epidemiol. 1992 Apr;45(4):373–6.

[19] Schmid AB, Bland JDP, Bhat MA, Bennett DLH. The relationship of nerve fibre pathology to sensory function in entrapment neuropathy. Brain J Neurol. 2014 Dec;137(Pt 12):3186–99.

[20] Han HY, Kim HM, Park SY, Kim M-W, Kim JM, Jang D-H. Clinical findings of asymptomatic carpal tunnel syndrome in patients with diabetes mellitus. Ann Rehabil Med. 2016 Jun;40(3):489–95.

[21] Gulliford MC, Latinovic R, Charlton J, Hughes RAC. Increased incidence of carpal tunnel syndrome up to 10 years before diagnosis of diabetes. Diabetes Care. 2006 Aug;29(8):1929–30.

[22] Oktayoglu P, Nas K, Kilinç F, Tasdemir N, Bozkurt M, Yildiz I. Assessment of the presence of carpal tunnel syndrome in patients with diabetes mellitus, hypothyroidism and acromegaly. J Clin Diagn Res. 2015 Jun;9(6):OC14–18.

[23] Shiri R, Pourmemari MH, Falah-Hassani K, Viikari-Juntura E. The effect of excess body mass on the risk of carpal tunnel syndrome: a meta-analysis of 58 studies. Obes Rev Off J Int Assoc Study Obes. 2015 Dec;16(12):1094–104.

[24] Meems M, Truijens S, Spek V, Visser LH, Pop VJM. Prevalence, course and determinants of carpal tunnel syndrome symptoms during pregnancy: a prospective study. BJOG Int J Obstet Gynaecol. 2015 Jul;122(8):1112–8.

[25] Nakagawa M, Sekijima Y, Yazaki M, Tojo K, Yoshinaga T, Doden T, Koyama J, Yanagisawa S, Ikeda S-I. Carpal tunnel syndrome: a common initial symptom of systemic wild-type ATTR (ATTRwt) amyloidosis. Amyloid Int J Exp Clin Investig Off J Int Soc Amyloid. 2016;23(1):58–63.

[26] Shiri R. Arthritis as a risk factor for carpal tunnel syndrome: a meta-analysis. Scand J Rheumatol. 2016 Oct;45(5):339–46.

[27] Atcheson SG, Ward JR, Lowe W. Concurrent medical disease in work-related carpal tunnel syndrome. Arch Intern Med. 1998 Jul 27;158(14):1506–12.

[28] Yamazaki T, Kawahara N, Arai K, Oyoshi K, Oshima M, Koike S, Miyauchi A, Hayasaka T, Saito T, Tsuruoka S. Utility of ultrasonography of the median nerve with a high-frequency probe for the diagnosis of dialysis-related carpal tunnel syndrome. Ther Apher Dial Off Peer-Rev J Int Soc Apher Jpn Soc Apher Jpn Soc Dial Ther. 2016 Oct;20(5):483–91.

[29] Derebery J. Work-related carpal tunnel syndrome: the facts and the myths. Clin Occup Environ Med. 2006;5(2):353–367, viii.

[30] Palmer KT, Harris EC, Coggon D. Carpal tunnel syndrome and its relation to occupation: a systematic literature review. Occup Med Oxf Engl. 2007 Jan;57(1):57–66.

[31] Perkins BA, Olaleye D, Bril V. Carpal tunnel syndrome in patients with diabetic polyneuropathy. Diabetes Care. 2002 Mar;25(3):565–9.

[32] Sax TW, Rosenbaum RB. Neuromuscular disorders in pregnancy. Muscle Nerve. 2006 Nov;34(5):559–71.

[33] Bouche P. Compression and entrapment neuropathies. Handb Clin Neurol. 2013;115:311–66.

[34] Bland JDP. Carpal tunnel syndrome. BMJ. 2007 Aug 18;335(7615):343–6.

[35] Borg K, Lindblom U. Diagnostic value of quantitative sensory testing (QST) in carpal tunnel syndrome. Acta Neurol Scand. 1988 Dec;78(6):537–41.

[36] Imai T, Wada T, Matsumoto H. Entrapment neuropathy of the palmar cutaneous branch of the median nerve in carpal tunnel syndrome. Clin Neurophysiol Off J Int Fed Clin Neurophysiol. 2004 Nov;115(11):2514–7.

[37] Buch N, Foucher G. Validity of clinical signs and provocative tests in carpal tunnel syndrome. Rev Chir Orthop Reparatrice Appar Mot. 1995;80(1):14–21.

[38] Padua L, Padua R, Aprile I, Pasqualetti P, Tonali P, Italian CTS Study Group. Carpal tunnel syndrome. Multiperspective follow-up of untreated carpal tunnel syndrome: a multicenter study. Neurology. 2001 Jun 12;56(11):1459–66.

[39] Katz JN, Larson MG, Sabra A, Krarup C, Stirrat CR, Sethi R, Eaton HM, Fossel AH, Liang MH. The carpal tunnel syndrome: diagnostic utility of the history and physical examination findings. Ann Intern Med. 1990 Mar 1;112(5):321–7.

[40] Wipperman J, Goerl K. Carpal tunnel syndrome: diagnosis and management. Am Fam Physician. 2016 Dec 15;94(12):993–9.

[41] Jablecki CK, Andary MT, Floeter MK, Miller RG, Quartly CA, Vennix MJ, Wilson JR, American Association of Electrodiagnostic Medicine, American Academy of Neurology, American Academy of Physical Medicine and Rehabilitation. Practice parameter: Electrodiagnostic studies in carpal tunnel syndrome. Report of the American Association of Electrodiagnostic Medicine, American Academy of Neurology, and the American Academy of Physical Medicine and Rehabilitation. Neurology. 2002 Jun 11;58(11):1589–92.

[42] Kutlar N, Bayrak AO, Bayrak İK, Canbaz S, Türker H. Diagnosing carpal tunnel syndrome with Doppler ultrasonography: a comparison of ultrasonographic measurements and electrophysiological severity. Neurol Res. 2017 Feb;39(2):126–32.

[43] Razek AAKA, Shabana AAE, El Saied TO, Alrefey N. Diffusion tensor imaging of mild-moderate carpal tunnel syndrome: correlation with nerve conduction study and clinical tests. Clin Rheumatol. 2016 Nov 3, 1–6, 10.1007/s10067-016-3463-y.

[44] Padua L, Coraci D, Erra C, Pazzaglia C, Paolasso I, Loreti C, Caliandro P, Hobson-Webb LD. Carpal tunnel syndrome: clinical features, diagnosis, and management. Lancet Neurol. 2016 Nov;15(12):1273–84.

[45] Ballestero-Pérez R, Plaza-Manzano G, Urraca-Gesto A, Romo-Romo F, Atín-Arratibel M de LÁ, Pecos-Martín D, Gallego-Izquierdo T, Romero-Franco N. Effectiveness of nerve gliding exercises on carpal tunnel syndrome: a systematic review. J Manipul Physiol Ther. 2017 Jan;40(1):50–9.

[46] Gerritsen AAM, de Vet HCW, Scholten RJPM, Bertelsmann FW, de Krom MCTFM, Bouter LM. Splinting vs. surgery in the treatment of carpal tunnel syndrome: a randomized controlled trial. JAMA. 2002 Sep 11;288(10):1245–51.

[47] Povlsen B, Bashir M, Wong F. Long-term result and patient reported outcome of wrist splint treatment for carpal tunnel syndrome. J Plast Surg Hand Surg. 2014 Jun;48(3):175–8.

[48] Mezian K, Bruthans J. Why do local corticosteroid injections work in carpal tunnel syndrome, but not in ulnar neuropathy at the elbow? Muscle Nerve. 2016 Aug;54(2):344.

[49] Huisstede BM, Randsdorp MS, Coert JH, Glerum S, van Middelkoop M, Koes BW. Carpal tunnel syndrome. Part II: effectiveness of surgical treatments—a systematic review. Arch Phys Med Rehabil. 2010 Jul;91(7):1005–24.

[50] Leblanc MR, Lalonde J, Lalonde DH. A detailed cost and efficiency analysis of performing carpal tunnel surgery in the main operating room versus the ambulatory setting in Canada. Hand N Y N. 2007 Dec;2(4):173–8.

[51] Chatterjee A, McCarthy JE, Montagne SA, Leong K, Kerrigan CL. A cost, profit, and efficiency analysis of performing carpal tunnel surgery in the operating room versus the clinic setting in the United States. Ann Plast Surg. 2011 Mar;66(3):245–8.

[52] Scholten RJPM, Mink van der Molen A, Uitdehaag BMJ, Bouter LM, de Vet HCW. Surgical treatment options for carpal tunnel syndrome. Cochrane Database Syst Rev. 2007 Oct 17;(4):CD003905.

[53] Thoma A, Veltri K, Haines T, Duku E. A meta-analysis of randomized controlled trials comparing endoscopic and open carpal tunnel decompression. Plast Reconstr Surg. 2004 Oct;114(5):1137–46.

[54] Zuo D, Zhou Z, Wang H, Liao Y, Zheng L, Hua Y, Cai Z. Endoscopic versus open carpal tunnel release for idiopathic carpal tunnel syndrome: a meta-analysis of randomized controlled trials. J Orthop Surg. 2015 Jan 28;10:12.

[55] Fernandes CH, Nakachima LR, Hirakawa CK, Gomes Dos Santos JB, Faloppa F. Carpal tunnel release using the paine retinaculotome inserted through a palmar incision. Hand N Y N. 2014 Mar;9(1):48–51.

[56] Jones NF, Ahn HC, Eo S. Revision surgery for persistent and recurrent carpal tunnel syndrome and for failed carpal tunnel release. Plast Reconstr Surg. 2012 Mar;129(3):683–92.

[57] Rasmussen MB, Deutch SR. Pronator teres syndrome is a rare but important cause of pain in the forearm. Ugeskr Laeger. 2016 Nov 14;178(46).

[58] Xing SG, Tang JB. Entrapment neuropathy of the wrist, forearm, and elbow. Clin Plast Surg. 2014 Jul;41(3):561–88.

[59] Olehnik WK, Manske PR, Szerzinski J. Median nerve compression in the proximal forearm. J Hand Surg. 1994 Jan;19(1):121–6.

[60] Lee HJ, Kim I, Hong JT, Kim MS. Early surgical treatment of pronator teres syndrome. J Korean Neurosurg Soc. 2014 May;55(5):296–9.

[61] Komaru Y, Inokuchi R. Anterior interosseous nerve syndrome. QJM Mon J Assoc Physicians. 2017 Jan 10;

[62] Ulrich D, Piatkowski A, Pallua N. Anterior interosseous nerve syndrome: retrospective analysis of 14 patients. Arch Orthop Trauma Surg. 2011 Nov;131(11):1561–5.

[63] Spinner M. The anterior interosseous-nerve syndrome, with special attention to its variations. J Bone Joint Surg Am. 1970 Jan;52(1):84–94.

[64] Damert H-G, Hoffmann R, Kraus A, Stowell RL, Lubahn J. Minimally invasive endoscopic decompression for anterior interosseous nerve syndrome: technical notes. J Hand Surg. 2013 Oct;38(10):2016–24.

[65] Feindel W, Stratford J. Cubital tunnel compression in tardy ulnar palsy. Can Med Assoc J. 1958 Mar 1;78(5):351–3.

[66] Bradshaw DY, Shefner JM. Ulnar neuropathy at the elbow. Neurol Clin. 1999 Aug;17(3): 447–461, v–vi.

[67] Apfelberg DB, Larson SJ. Dynamic anatomy of the ulnar nerve at the elbow. Plast Reconstr Surg. 1973 Jan;51(1):79–81.

[68] Mondelli M, Giannini F, Ballerini M, Ginanneschi F, Martorelli E. Incidence of ulnar neuropathy at the elbow in the province of Siena (Italy). J Neurol Sci. 2005 Jul 15;234(1–2):5–10.

[69] Contreras MG, Warner MA, Charboneau WJ, Cahill DR. Anatomy of the ulnar nerve at the elbow: potential relationship of acute ulnar neuropathy to gender differences. Clin Anat N Y N. 1998;11(6):372–8.

[70] Cutts S. Cubital tunnel syndrome. Postgrad Med J. 2007 Jan;83(975):28–31.

[71] Latinovic R, Gulliford MC, Hughes R A. Incidence of common compressive neuropathies in primary care. J Neurol Neurosurg Psychiatry. 2006 Feb;77(2):263–5.

[72] Richardson JK, Green DF, Jamieson SC, Valentin FC. Gender, body mass and age as risk factors for ulnar mononeuropathy at the elbow. Muscle Nerve. 2001 Apr;24(4):551–4.

[73] Descatha A, Leclerc A, Chastang. J-F, Roquelaure Y. Incidence of ulnar nerve entrapment at the elbow in repetitive work. Scand J Work Envrion Health. 2004;30:234–6.

[74] van Rijn RM, Huisstede BMA, Koes BW, Burdorf A. Associations between work-related factors and specific disorders at the elbow: a systematic literature review. Rheumatol Oxf Engl. 2009 May;48(5):528–36.

[75] Patterson JMM, Jaggars MM, Boyer MI. Ulnar and median nerve palsy in long-distance cyclists. A prospective study. Am J Sports Med. 2003 Aug;31(4):585–9.

[76] Akuthota V, Plastaras C, Lindberg K, Tobey J, Press J, Garvan C. The effect of long-distance bicycling on ulnar and median nerves: an electrophysiologic evaluation of cyclist palsy. Am J Sports Med. 2005 Aug;33(8):1224–30.

[77] Mondelli M, Grippo A, Mariani M, Baldasseroni A, Ansuini R, Ballerini M, Bandinelli C, Graziani M, Luongo F, Mancini R, Manescalchi P, Pellegrini S, Sgarrella C, Giannini F. Carpal tunnel syndrome and ulnar neuropathy at the elbow in floor cleaners. Neurophysiol Clin Clin Neurophysiol. 2006 Aug;36(4):245–53.

[78] Tagliafico A, Resmini E, Nizzo R, Derchi LE, Minuto F, Giusti M, Martinoli C, Ferone D. The pathology of the ulnar nerve in acromegaly. Eur J Endocrinol. 2008 Oct;159(4):369–73.

[79] Balagtas-Balmaseda OM, Grabois M, Balmaseda PF, Lidsky MD. Cubital tunnel syndrome in rheumatoid arthritis. Arch Phys Med Rehabil. 1983 Apr;64(4):163–6.

[80] Shinohara T, Tatebe M, Okui N, Yamamoto M, Kurimoto S, Hirata H. Cubital tunnel syndrome caused by amyloid elbow arthropathy in long-term hemodialysis patients: report of 4 cases. J Hand Surg. 2011 Oct;36(10):1640–3.

[81] Harder K, Lukschu S, Dunda SE, Krapohl BD. Results after simple decompression of the ulnar nerve in cubital tunnel syndrome. GMS Interdiscip Plast Reconstr Surg DGPW. 2015;4:Doc19.

[82] Nathan PA, Istvan JA, Meadows KD. Intermediate and long-term outcomes following simple decompression of the ulnar nerve at the elbow. Chir Main. 2005 Feb;24(1):29–34.

[83] Seror P, Nathan PA. Relative frequency of nerve conduction abnormalities at carpal tunnel and cubital tunnel in France and the United States: importance of silent neuropathies and role of ulnar neuropathy after unsuccessful carpal tunnel syndrome release. Ann Chir Main Memb Superieur Organe Off Soc Chir Main Ann Hand Up Limb Surg. 1993;12(4):281–5.

[84] Mallette P, Zhao M, Zurakowski D, Ring D. Muscle atrophy at diagnosis of carpal and cubital tunnel syndrome. J Hand Surg. 2007 Aug;32(6):855–8.

[85] Hayashi Y, Kojima T, Kohno T. A case of cubital tunnel syndrome caused by the snapping of the medial head of the triceps brachii muscle. J Hand Surg. 1984 Jan;9A(1):96–9.

[86] Morgenstein A, Lourie G, Miller B. Anconeus epitrochlearis muscle causing dynamic cubital tunnel syndrome: a case series. J Hand Surg Eur Vol. 2016 Feb;41(2):227–9.

[87] Nagle DJ, Patel RM, Paisley S. Endoscopic detection of compressing fascial bands around the ulnar nerve within the FCU. Hand N Y N. 2012 Mar;7(1):103–7.

[88] Earp BE, Floyd WE, Louie D, Koris M, Protomastro P. Ulnar nerve entrapment at the wrist. J Am Acad Orthop Surg. 2014 Nov;22(11):699–706.

[89] Trehan SK, Parziale JR, Akelman E. Cubital tunnel syndrome: diagnosis and management. Med Health R I. 2012 Nov;95(11):349–52.

[90] Prielipp RC, Morell RC, Butterworth J. Ulnar nerve injury and perioperative arm positioning. Anesthesiol Clin N Am. 2002 Sep;20(3):589–603.

[91] Asamoto S, Böker D-K, Jödicke A. Surgical treatment for ulnar nerve entrapment at the elbow. Neurol Med Chir (Tokyo). 2005 May;45(5):240–245.

[92] Bacle G, Marteau E, Freslon M, Desmoineaux P, Saint-Cast Y, Lancigu R, Kerjean Y, Vernet E, Fournier J, Corcia P, Le Nen D, Rabarin F, Laulan J. Cubital tunnel syndrome: comparative results of a multicenter study of 4 surgical techniques with a mean follow-up of 92 months. Orthop Traumatol Surg Res. 2014 Jun;100(4 Suppl):S205–208.

[93] Keiner D, Gaab MR, Schroeder HWS, Oertel J. Comparison of the long-term results of anterior transposition of the ulnar nerve or simple decompression in the treatment of cubital tunnel syndrome—a prospective study. Acta Neurochir (Wien). 2009 Apr;151(4):311–315; discussion 316.

[94] Sousa M, Aido R, Trigueiros M, Lemos R, Silva C. Cubital compressive neuropathy in the elbow: in situ neurolysis versus anterior transposition - comparative study. Rev Bras Ortop. 2014 Dec;49(6):647–52.

[95] Chen H, Ou S, Liu G, Fei J, Zhao G, Wu L, Pan J. Clinical efficacy of simple decompression versus anterior transposition of the ulnar nerve for the treatment of cubital tunnel syndrome: a meta-analysis. Clin Neurol Neurosurg. 2014 Nov;126:150–5.

[96] Macadam SA, Gandhi R, Bezuhly M, Lefaivre KA. Simple decompression versus anterior subcutaneous and submuscular transposition of the ulnar nerve for cubital tunnel syndrome: a meta-analysis. J Hand Surg. 2008 Oct;33(8):1314.e1–12.

[97] Zlowodzki M, Chan S, Bhandari M, Kalliainen L, Schubert W. Anterior transposition compared with simple decompression for treatment of cubital tunnel syndrome. A meta-analysis of randomized, controlled trials. J Bone Joint Surg Am. 2007 Dec;89(12):2591–8.

[98] Biggs M, Curtis JA. Randomized, prospective study comparing ulnar neurolysis in situ with submuscular transposition. Neurosurgery. 2006 Feb;58(2):296–304–304.

[99] Bartels RHMA, Verhagen WIM, van der Wilt GJ, Meulstee J, van Rossum LGM, Grotenhuis JA. Prospective randomized controlled study comparing simple decompression versus anterior subcutaneous transposition for idiopathic neuropathy of the ulnar nerve at the elbow: Part 1. Neurosurgery. 2005 Mar;56(3):522–530–530.

[100] Schmidt S, Kleist Welch-Guerra W, Matthes M, Baldauf J, Schminke U, Schroeder HWS. Endoscopic vs open decompression of the ulnar nerve in cubital tunnel syndrome: a prospective randomized double-blind study. Neurosurgery. 2015 Dec;77(6):960–971.

[101] Sener S, Menovsky T, Kloet A. Open ulnar nerve decompression using small incision and alternate positioning. Neurosurgery. 2014 Feb;74(2):E230–232.

[102] Manske PR, Johnston R, Pruitt DL, Strecker WB. Ulnar nerve decompression at the cubital tunnel. Clin Orthop. 1992 Jan;(274):231–7.

[103] Bartels RHMA, Termeer EH, van der Wilt GJ, van Rossum LGM, Meulstee J, Verhagen WIM, Grotenhuis JA. Simple decompression or anterior subcutaneous transposition for ulnar neuropathy at the elbow: a cost-minimization analysis—Part 2. Neurosurgery. 2005 Mar;56(3):531–536–536.

[104] Dellon AL, MacKinnon SE. Injury to the medial antebrachial cutaneous nerve during cubital tunnel surgery. J Hand Surg Edinb Scotl. 1985 Feb;10(1):33–6.

[105] Lowe JB, Maggi SP, Mackinnon SE. The position of crossing branches of the medial antebrachial cutaneous nerve during cubital tunnel surgery in humans. Plast Reconstr Surg. 2004 Sep 1;114(3):692–6.

[106] Ruchelsman DE, Lee SK, Posner MA. Failed surgery for ulnar nerve compression at the elbow. Hand Clin. 2007 Aug;23(3):359–371, vi–vii.

[107] Yoshida A, Okutsu I, Hamanaka I. Endoscopic anatomical nerve observation and minimally invasive management of cubital tunnel syndrome. J Hand Surg Eur Vol. 2009 Feb;34(1):115–20.

[108] Dützmann S, Martin KD, Sobottka S, Marquardt G, Schackert G, Seifert V, Krishnan KG. Open vs retractor-endoscopic in situ decompression of the ulnar nerve in cubital tunnel syndrome: a retrospective cohort study. Neurosurgery. 2013 Apr;72(4):605–616–616.

[109] Tang P, Hoellwarth JS, Chauhan A. Recurrent cubital tunnel syndrome: a critical analysis review. JBJS Rev. 2016 Mar 8;4(3).

[110] Ehsan A, Hanel DP. Recurrent or persistent cubital tunnel syndrome. J Hand Surg. 2012 Sep;37(9):1910–2.

[111] Campbell WW, Pridgeon RM, Sahni KS. Short segment incremental studies in the evaluation of ulnar neuropathy at the elbow. Muscle Nerve. 1992 Sep;15(9):1050–4.

[112] Kanakamedala RV, Simons DG, Porter RW, Zucker RS. Ulnar nerve entrapment at the elbow localized by short segment stimulation. Arch Phys Med Rehabil. 1988 Nov;69(11):959–63.

[113] Raynor EM, Shefner JM, Preston DC, Logigian EL. Sensory and mixed nerve conduction studies in the evaluation of ulnar neuropathy at the elbow. Muscle Nerve. 1994 Jul;17(7):785–92.

[114] Cho Y-J, Cho S-M, Sheen S-H, Choi J-H, Huh D-H, Song J-H. Simple decompression of the ulnar nerve for cubital tunnel syndrome. J Korean Neurosurg Soc. 2007 Nov;42(5):382–7.

[115] Kothari MJ. Ulnar neuropathy at the wrist. Neurol Clin. 1999 Aug;17(3):463–476, vi.

[116] Bouche P, Séror R, Psimaras D, Séror P, Ebelin M. Entrapment neuropathies involving the ulnar nerve at the wrist (and into the hand) and the peroneal nerve. Rev Neurol (Paris). 2008 Dec;164(12):1073–6.

[117] Kerasnoudis A, Ntasiou P, Ntasiou E. Prognostic value of nerve ultrasound and electrophysiological findings in Saturday night palsy. J Neuroimaging Off J Am Soc Neuroimaging. 2016 Dec 5;

[118] Reichert P, Wnukiewicz W, Witkowski J, Bocheńska A, Mizia S, Gosk J, Zimmer K. Causes of secondary radial nerve palsy and results of treatment. Med Sci Monit Int Med J Exp Clin Res. 2016 Feb 19;22:554–62.

[119] Naam NH, Nemani S. Radial tunnel syndrome. Orthop Clin North Am. 2012 Oct;43(4): 529–36.

[120] Urch EY, Model Z, Wolfe SW, Lee SK. Anatomical study of the surgical approaches to the radial tunnel. J Hand Surg. 2015 Jul;40(7):1416–20.

[121] Konjengbam M, Elangbam J. Radial nerve in the radial tunnel: anatomic sites of entrapment neuropathy. Clin Anat N Y N. 2004 Jan;17(1):21–5.

[122] Stanley J. Radial tunnel syndrome: a surgeon's perspective. J Hand Ther Off J Am Soc Hand Ther. 2006 Jun;19(2):180–4.

[123] Ke M-J, Chen L-C, Chou Y-C, Li T-Y, Chu H-Y, Tsai C-K, Wu Y-T. The dose-dependent efficiency of radial shock wave therapy for patients with carpal tunnel syndrome: a prospective, randomized, single-blind, placebo-controlled trial. Sci Rep. 2016 Dec 2;6:38344.

[124] Moradi A, Ebrahimzadeh MH, Jupiter JB. Radial tunnel syndrome, diagnostic and treatment dilemma. Arch Bone J Surg. 2015 Jul;3(3):156–62.

[125] Sun Y, Tong P-J, Li X-J. Entrapment syndrome of posterior interosseous nerve caused by elbow cyst: 5 cases reports. Zhongguo Gu Shang China J Orthop Traumatol. 2013 Nov;26(11):949–52.

[126] Quignon R, Marteau E, Penaud A, Corcia P, Laulan J. Posterior interosseous nerve palsy. A series of 18 cases and literature review. Chir Main. 2012 Feb;31(1):18–23.

[127] Borman P, Tuncay F, Ulusoy G, Koçer U. Posterior interosseous nerve syndrome due to lipoma. Neurophysiol Clin Clin Neurophysiol. 2010 Jun;40(3):189–91.

[128] Wu P, Yang JY, Yu C. Surgical treatment of spontaneous posterior interosseous nerve palsy with hourglass-like constriction. J Hand Surg Eur Vol. 2015 Jul;40(6):646–7.

[129] Ehrlich W, Dellon AL, Mackinnon SE. Classical article: Cheiralgia paresthetica (entrapment of the radial nerve). A translation in condensed form of Robert Wartenberg's original article published in 1932. J Hand Surg. 1986 Mar;11(2):196–9.

[130] Dorfman LJ, Jayaram AR. Handcuff-related cheiralgia paresthetica. Neurology. 1979 Jun;29(6):908–9.

[131] Stahl S, Kaufman T. Cheiralgia paresthetica—entrapment of the superficial branch of the radial nerve: a report of 15 cases. Eur J Plast Surg. 1997;20(2):57–9.

[132] Dellon AL, Mackinnon SE. Radial sensory nerve entrapment. Arch Neurol. 1986 Aug; 43(8):833–5.

[133] Spies CK, Müller LP, Oppermann J, Neiss WF, Hahn P, Unglaub F. Surgical decompression of the superficial radial nerve: Wartenberg syndrome. Oper Orthopadie Traumatol. 2016 Apr;28(2):145–52.

[134] Kumar A, Sharma A, Singh P. Anatomical study of the suprascapular notch: quantitative analysis and clinical considerations for suprascapular nerve entrapment. Singapore Med J. 2014 Jan;55(1):41–4.

[135] Hill LJN, Jelsing EJ, Terry MJ, Strommen JA. Evaluation, treatment, and outcomes of suprascapular neuropathy: a 5-year review. PM R. 2014 Sep;6(9):774–80.

[136] Yamakado K. Quantification of the learning curve for arthroscopic suprascapular nerve decompression: an evaluation of 300 cases. Arthrosc J Arthrosc Relat Surg Off Publ Arthrosc Assoc N Am Int Arthrosc Assoc. 2015 Feb;31(2):191–6.

[137] Ferrante MA, Ferrante ND. The thoracic outlet syndromes: part 1. Overview of the thoracic outlet syndromes and review of true neurogenic thoracic outlet syndrome. Muscle Nerve. 2016 Dec 22;

[138] Laulan J. Thoracic outlet syndromes. The so-called "neurogenic types." Hand Surg Rehabil. 2016 Jun;35(3):155–64.

[139] Marty F-L, Corcia P, Alexandre J, Laulan J. True neurological thoracic outlet syndrome. Retrospective study of 30 consecutive cases. Chir Main. 2012 Oct;31(5):244–9.

[140] Mackinnon SE, Novak CB. Thoracic outlet syndrome. Curr Probl Surg. 2002 Nov; 39(11):1070–145.

[141] Hooper TL, Denton J, McGalliard MK, Brismée J-M, Sizer PS. Thoracic outlet syndrome: a controversial clinical condition. Part 1: anatomy, and clinical examination/diagnosis. J Man Manip Ther. 2010 Jun;18(2):74–83.

[142] Machleder HI. The treatment of thoracic outlet syndrome: a comparison of different operations. J Vasc Surg. 1990 Aug;12(2):220–1.

[143] Abdel Ghany W, Nada MA, Toubar AF, Desoky AE, Ibrahim H, Nassef MA, Mahran MG. Modified interscalene approach for resection of symptomatic cervical rib: anatomical review and clinical study. World Neurosurg. 2016 Oct 27;

[144] Seror P, Seror R. Meralgia paresthetica: clinical and electrophysiological diagnosis in 120 cases. Muscle Nerve. 2006 May;33(5):650–4.

[145] van Slobbe AM, Bohnen AM, Bernsen RMD, Koes BW, Bierma-Zeinstra SMA. Incidence rates and determinants in meralgia paresthetica in general practice. J Neurol. 2004 Mar;251(3):294–7.

[146] Craig A. Entrapment neuropathies of the lower extremity. P M R. 2013 May;5(5 Suppl): S31–40.

[147] Parisi TJ, Mandrekar J, Dyck PJB, Klein CJ. Meralgia paresthetica: relation to obesity, advanced age, and diabetes mellitus. Neurology. 2011 Oct 18;77(16):1538–42.

[148] Cheatham SW, Kolber MJ, Salamh PA. Meralgia paresthetica: a review of the literature. Int J Sports Phys Ther. 2013 Dec;8(6):883–93.

[149] Lee S-H, Shin K-J, Gil Y-C, Ha T-J, Koh K-S, Song W-C. Anatomy of the lateral femoral cutaneous nerve relevant to clinical findings in meralgia paresthetica. Muscle Nerve. 2016 Aug 20;

[150] Mondelli M, Rossi S, Romano C. Body mass index in meralgia paresthetica: a case-control study. Acta Neurol Scand. 2007 Aug;116(2):118–23.

[151] Van Diver T, Camann W. Meralgia paresthetica in the parturient. Int J Obstet Anesth. 1995 Apr;4(2):109–12.

[152] Pauwels A, Amarenco P, Chazouillères O, Pigot F, Calmus Y, Lévy VG. Unusual and unknown complication of ascites: meralgia paresthetica. Gastroenterol Clin Biol. 1990;14(3):295.

[153] Moucharafieh R, Wehbe J, Maalouf G. Meralgia paresthetica: a result of tight new trendy low cut trousers ('taille basse'). Int J Surg Lond Engl. 2008 Apr;6(2):164–8.

[154] Chauhan G, Gupta K, Nayar P. Meralgia paresthetica after total hip arthroplasty in supine position. Saudi J Anaesth. 2013 Jan;7(1):105–6.

[155] Shin HJ, Kim YH, Lee HW. Meralgia paresthetica-like symptoms following epidural analgesia after total knee arthroplasty. Acta Anaesthesiol Scand. 2014 Nov;58(10):1276–9.

[156] Suh DH, Kim DH, Park JW, Park BK. Sonographic and electrophysiologic findings in patients with meralgia paresthetica. Clin Neurophysiol Off J Int Fed Clin Neurophysiol. 2013 Jul;124(7):1460–4.

[157] Aravindakannan T, Wilder-Smith EP. High-resolution ultrasonography in the assessment of meralgia paresthetica. Muscle Nerve. 2012 Mar;45(3):434–5.

[158] Klauser AS, Abd Ellah MMH, Halpern EJ, Sporer I, Martinoli C, Tagliafico A, Sojer M, Taljanovic MS, Jaschke WR. Meralgia paraesthetica: ultrasound-guided injection at multiple levels with 12-month follow-up. Eur Radiol. 2016 Mar;26(3):764–70.

[159] de Ruiter GCW, Wurzer JAL, Kloet A. Decision making in the surgical treatment of meralgia paresthetica: neurolysis versus neurectomy. Acta Neurochir (Wien). 2012 Oct;154(10):1765–72.

[160] Tzika M, Paraskevas G, Natsis K. Entrapment of the superficial peroneal nerve: an anatomical insight. J Am Podiatr Med Assoc. 2015 Mar;105(2):150–9.

[161] Bregman PJ, Schuenke M. Current diagnosis and treatment of superficial fibular nerve injuries and entrapment. Clin Podiatr Med Surg. 2016 Apr;33(2):243–54.

[162] Poage C, Roth C, Scott B. Peroneal nerve palsy: evaluation and management. J Am Acad Orthop Surg. 2016 Jan;24(1):1–10.

[163] Koller RL, Blank NK. Strawberry pickers' palsy. Arch Neurol. 1980 May;37(5):320.

[164] Stewart JD. Foot drop: where, why and what to do? Pract Neurol. 2008 Jun;8(3):158–69.

[165] Iwamoto N, Isu T, Chiba Y, Kim K, Morimoto D, Yamazaki K, Isobe M. Clinical feathers and treatment of peroneal nerve entrapment neuropathy. No Shinkei Geka. 2015 Apr;43(4):309–16.

[166] Morimoto D, Isu T, Kim K, Imai T, Yamazaki K, Matsumoto R, Isobe M. Surgical treatment of superior cluneal nerve entrapment neuropathy. J Neurosurg Spine. 2013 Jul;19(1):71–5.

[167] Logullo F, Ganino C, Lupidi F, Perozzi C, Di Bella P, Provinciali L. Anterior tarsal tunnel syndrome: a misunderstood and a misleading entrapment neuropathy. Neurol Sci Off J Ital Neurol Soc Ital Soc Clin Neurophysiol. 2014 May;35(5):773–5.

[168] DiDomenico LA, Masternick EB. Anterior tarsal tunnel syndrome. Clin Podiatr Med Surg. 2006 Jul;23(3):611–20.

[169] Pfeiffer WH, Cracchiolo A. Clinical results after tarsal tunnel decompression. J Bone Joint Surg Am. 1994 Aug;76(8):1222–30.

[170] Kawakatsu M, Ishiko T, Sumiya M. Tarsal tunnel syndrome due to three different types of ganglion during a 12-year period: a case report. J Foot Ankle Surg Off Publ Am Coll Foot Ankle Surg. 2017 Jan 7;

[171] Pomeroy G, Wilton J, Anthony S. Entrapment neuropathy about the foot and ankle: an update. J Am Acad Orthop Surg. 2015 Jan;23(1):58–66.

[172] Pardal-Fernández JM. Bilateral tarsal tunnel syndrome due to synovitis. Combined diagnostic contribution made by ultrasound and electrophysiology. Rev Neurol. 2013 Jan 16;56(2):124–5.

[173] Coraci D, Ioppolo F, Di Sante L, Santilli V, Padua L. Ultrasound in tarsal tunnel syndrome: correct diagnosis for appropriate treatment. Muscle Nerve. 2016 Dec;54(6):1148–9.

[174] Wong GNL, Tan TJ. MR imaging as a problem solving tool in posterior ankle pain: a review. Eur J Radiol. 2016 Dec;85(12):2238–56.

[175] Yassin M, Garti A, Weissbrot M, Heller E, Robinson D. Treatment of anterior tarsal tunnel syndrome through an endoscopic or open technique. Foot Edinb Scotl. 2015 Sep;25(3):148–51.

[176] Reichert P, Zimmer K, Wnukiewicz W, Kuliński S, Mazurek P, Gosk J. Results of surgical treatment of tarsal tunnel syndrome. Foot Ankle Surg Off J Eur Soc Foot Ankle Surg. 2015 Mar;21(1):26–9.

[177] Cass S. Upper extremity nerve entrapment syndromes in sports: an update. Curr Sports Med Rep. 2014 Feb;13(1):16–21.

[178] Siddiq MAB, Hossain MS, Uddin MM, Jahan I, Khasru MR, Haider NM, Rasker JJ. Piriformis syndrome: a case series of 31 Bangladeshi people with literature review. Eur J Orthop Surg Traumatol Orthop Traumatol. 2016 Sep 19;

[179] Carro LP, Hernando MF, Cerezal L, Navarro IS, Fernandez AA, Castillo AO. Deep gluteal space problems: piriformis syndrome, ischiofemoral impingement and sciatic nerve release. Muscles Ligaments Tendons J. 2016 Sep;6(3):384–96.

[180] Fishman LM, Wilkins AN, Rosner B. Electrophysiologically identified piriformis syndrome is successfully treated with incobotulinum toxin A and physical therapy. Muscle Nerve. 2016 Dec 9;

[181] Park M-S, Yoon S-J, Jung S-Y, Kim S-H. Clinical results of endoscopic sciatic nerve decompression for deep gluteal syndrome: mean 2-year follow-up. BMC Musculoskelet Disord. 2016 May 20;17:218.

[182] Khoder W, Hale D. Pudendal neuralgia. Obstet Gynecol Clin North Am. 2014 Sep;41(3):443–52.

[183] Ploteau S, Cardaillac C, Perrouin-Verbe M-A, Riant T, Labat J-J. Pudendal neuralgia due to pudendal nerve entrapment: warning signs observed in two cases and review of the literature. Pain Physician. 2016 Mar;19(3):E449–454.

[184] Andersen KV, Bovim G. Impotence and nerve entrapment in long distance amateur cyclists. Acta Neurol Scand. 1997 Apr;95(4):233–40.

[185] Örmeci B, Avcı E, Kaspar Ç, Terim ÖE, Erdoğru T, Öge AE. A novel electrophysiological method in the diagnosis of pudendal neuropathy: position-related changes in pudendal sensory evoked potentials. Urology. 2017 Jan;99:288.e1–288.e7.

[186] Labat JJ, Riant T, Lassaux A, Rioult B, Rabischong B, Khalfallah M, Volteau C, Leroi A-M, Ploteau S. Adding corticosteroids to the pudendal nerve block for pudendal neuralgia: a randomised, double-blind, controlled trial. BJOG Int J Obstet Gynaecol. 2017 Jan;124(2):251–60.

[187] Pérez-López FR, Hita-Contreras F. Management of pudendal neuralgia. Climacteric J Int Menopause Soc. 2014 Dec;17(6):654–6.

Chapter 3

Nerve Transfers in the Treatment of Peripheral Nerve Injuries

Vicente Vanaclocha-Vanaclocha,

Jose María Ortiz-Criado, Nieves Sáiz-Sapena and

Nieves Vanaclocha

Additional information is available at the end of the chapter

http://dx.doi.org/10.5772/67948

Abstract

Successful re-innervation of proximal limb peripheral nerve injuries is rare. Axons regenerate at ~1 mm/day, reaching hand muscles by 24 months, finding them atrophied and fibrosed. Peripheral nerve injury repair is often delayed waiting for spontaneous recovery. This waiting time should not be longer than 6 months as after 18 months reinnervation will not achieve effective muscular function. When spontaneous recovery is impossible, referral too late or damage too severe, other options like a transfer from a nearby healthy nerve to the injured one must be considered. They are very successful, and the deficit in the donor site is usually minimal. The most common nerve transfers are a branch of the spinal nerve to the trapezius muscle to the suprascapular nerve, a branch of the long head of the triceps to the axillary nerve, a fascicle of the ulnar nerve to the motor branch of the biceps muscle, two branches of the median nerve to the posterior interosseous nerve and the anterior interosseous nerve to the ulnar nerve. There are many more options that can suit particular cases. Introduced in brachial plexus injury repair, they are now also applied to lower limb, to stroke and to some spinal cord injuries.

Keywords: peripheral nerve injuries, brachial plexus injuries, nerve transfers, nerve regeneration, supercharge end-to-side, nerve repair, nerve graft

1. Introduction

Peripheral nerve injuries can represent a serious problem, particularly when involving the brachial plexus. They usually induce devastating consequences and affect young people [1]. The main cause is traffic accidents.

Motor endplate degeneration starts just after a motor nerve is injured. The growth speed of the regenerating axons is ~1 mm/day [2]. The time needed for the regeneration will be proportional the distance between the injury site and the muscle endplates [3, 4]. Proximal limb peripheral nerve injuries pose severe difficulties to a successful re-innervation because of the long distance regenerating axons have to cover [3], taking up to 24 months to reach hands or feet [2, 5, 6], often finding them atrophied and substituted by fat and fibrous tissue [7–9].

Peripheral nerve injury repair is often delayed waiting for spontaneous recovery. This waiting time should be no longer than 6 months [10] as after 18 months successful re-innervation will not achieve effective muscular function [11–14].

Nerve grafts are often required [15, 16]. Unfortunately, their use is associated with worse results than direct repair [17], particularly in grafts longer than 7.2 cm [1, 18]. This occurs because the regenerating nerve fibres must cross two anastomoses instead of just one. The second anastomose site is reached much later with more fibrosis hindering the nerve fibre growth to cross it [19]. Another reason for this is that only autologous sensory nerves are used as grafts to minimize donor site deficits [20]. It has been shown that using sensory nerve grafts to repair motor nerve defects is associated with worse regeneration results than if motor nerves are repaired with motor nerve grafts [21].

Direct repair is not always possible: when there is no chance of spontaneous recovery (i.e. root avulsion), the referral is too late, the damage is too severe, the scarring at the injury site is significant, presence of large neuromas in continuity or in multilevel nerve injuries [22]. In these cases, a nerve transfer from a nearby healthy nerve is a superior option. A healthy nerve is transected and coapted to a nearby injured one. This transforms a proximal nerve injury into a distal one near the motor endplates, reducing the time required for re-innervation [10, 13]. Unfortunately, the functional improvement is at the expense of reducing the number of functioning nerves [23]. Obviously, the function lost must be less essential than the one we expect to recover [24]. They yield better results than direct nerve repairs, particularly in proximal limb injuries and when a nerve graft is needed [9, 25, 26]. With good surgical planning the deficits induced on transecting the donor nerve are minimal [13, 24].

Distal nerve transfers in proximal nerve injuries reduce the time muscles are denervated, thus improving the outcomes [24, 27]. Nevertheless, the best results are obtained with simultaneous proximal nerve injury repair and distal nerve transfers [9] as this combination re-innervates more muscles [9].

Nerve transfers offer better results than tendon transfers [28]. They preserve the original anatomical situation of muscles and tendons, allowing a much better physiological function [11, 12, 14, 22, 29, 30]. Moreover, nerve transfers can re-innervate more than one muscle and thus recover more than one function [11, 29, 31, 32] while tendon transfers are limited to a single one [28]. Although initially introduced to repair peripheral nerve injuries, they are lately also being used in some cases of spinal cord injury (SCI) or stroke [33].

The mismatch between donor and recipient nerves is common (the first has fewer motor axons than the second). This is seldom a problem because a 20% of motor axons are enough to re-innervate the whole muscle [34, 35].

This means that each motor axon can increase up to five times the amount of muscle fibres it innervates [36].

Thus, it is possible to re-innervate a big nerve with a smaller branch but it comes at the price of courser movements [34]. A reduced number of axons can successfully re-innervate a muscle provided it is done on time [37].

Nerve transfer indications: irrecoverable proximal nerve injury (i.e. root avulsion) [38], if a long nerve graft is required (over 7.2 cm results are dismal) [1, 18], long distance between injury site and motor endplates [5], late presentation [8], very wide injury and dense scar tissue [10].

Nerve transfer contraindications: existence of a better option, time since injury over 18 months [5] and motor donor strength below Medical Research Council (MRC) grade M4 [12].

Ideal donor nerve: purely motor or sensory, containing enough axons to re-innervate the recipient muscles, its diameter is similar to the recipient nerve, requires no nerve graft, innervates expendable muscles [24, 39–41], and has synergistic function with the recipient nerve [10, 12, 13, 22, 27, 33]. The functional loss created on taking the donor should be less important than the functional recovery expected on re-innervating the recipient [24, 42]. Post-operative rehabilitation is easier when donor and recipient nerves have a synergistic action [22, 27, 32]. In the case of antagonistic action, much more post-operative re-education will be needed to recover the same amount of function [27]. The recovery of muscle power depends on the amount of motor axons provided by the donor nerve [13, 27, 35] and on the time elapsed until re-innervation happens [13, 29]. It is known that a muscle generates a normal power until about 80% of the motor axons are lost, but afterwards, there is a sharp loss [34]. Thus, it is crucial to be above this 20% [19, 27].

To allow a tension-free repair, essential for a successful recovery, the donor nerve must be transected as distally as possible and the recipient nerve as proximally as feasible [14, 15, 31].

Nerve transfer advantages: shorter distance between donor healthy nerve and denervated muscle endplates, safe supply of viable axons, usually no nerve graft needed, selection of pure motor or sensory axons, possibility to recover more than one function and no scar in the surgical field [13, 22, 27].

Nerve transfer disadvantages: a function has to be sacrificed, donor site morbidity, donor and recipient muscle co-contraction and possibility of previous donor nerve injury [27].

Donor site morbidity: shown in the spinal accessory (SAN) to suprascapular nerve (SSN) transfer. The weakness of the middle and lower part of the trapezius muscle (donor site) induces mild scapular winging in the case of good recovery of shoulder external rotation.

Co-contraction: it can be useful in the case of synergistic action between donor and recipient muscles. This is the case in the SAN to SSN transfer as the trapezius is synergistic with the supraspinatus and infraspinatus action. In most other cases it is an inconvenience [22, 24, 43, 44]. For example, in the Oberlin procedure (a fascicle of the ulnar nerve (UN) is transferred to the biceps muscle (BM) nerve branch), there is a tendency for finger flexion when attempting elbow flexion [45]. This is also the case of medial pectoral nerve (MPN) to musculocutaneous nerve (MCN) transfer in which elbow flexion is associated with shoulder adduction [45]. A more dramatic example is phrenic nerve (PHN) to radial nerve (RAN) transfer as patients must take a deep breath before attempting hand extension [44, 46]. In some cases, it is so serious that it can make the procedure useless (i.e. contralateral C_7 nerve root transfer) [43].

Previous damage to the donor nerve: it is always a possibility as muscle weakness is only clinically evident when at least 50% of the motor axons are lost [45]. It may explain the variability in the clinical outcomes [45, 47].

An adequate *post-operative rehabilitation* is a vital element in the final outcome and relies on cerebral plasticity in which neurons are assigned to new tasks [4, 48].

Initially, motor nerve transfers were the main concern. Over time it became obvious that sensory recovery is also essential as it allows a better motor control and avoids trophic ulcers [49–52]. This leads to the introduction of sensory nerve transfers [50, 52].

Nerve transfers were first used in the upper limbs but with time have also been applied to the lower limbs.

1.1. History

Balance in 1903 was the first to report a nerve transfer (SAN to facial nerve) [53] but it was Tuttle in 1913 the first to use them to repair brachial plexus injuries [54]. Vulpius and Stoffel [55] in 1920 described the use of the MPN as donor nerve [56]. Harris in 1921 reported the RAN to median nerve (MN) transfer [57]. Förster [58] in 1929 transferred the thoracodorsal (TDN) and subscapular nerves to the axillary nerve (AXN). Lurje in 1948 transferred the pectoral and TDN nerves to the MCN [59]. Seddon in 1963 described the use of the intercostal nerves (ICNs) as donor nerves [16]. Samardzic et al. in 1980 performed the first double nerve transfer, pectoral to AXN and TDN to MCN [60]. Bedeschi et al. in 1984 introduced the sensory nerve transfers [61]. Novak and Mackinnon in 1991 reported the pronator *quadratus* (PQ) to UN motor fascicle transfer in proximal ulnar nerve lesions [62]. Brandt and Mackinnon [63] in 1993 reported their experience with pectoral to MCN transfer and Oberlin et al. [64] in 1994 reported the UN to biceps muscle nerve (BMN) transfer. Ever since, many nerve transfers have been reported. Experience has settled the indications and outcomes of each of them. Mastering all of them allows adaptation to all circumstances as the ideal donor is not always available.

1.2. Types of nerve transfers

Attending to the nerves involved they can be classified as motor or sensory. The *motor transfers* aim to recover movement and to avoid subluxation (common in a denervated shoulder) [24].

http://dx.doi.org/10.5772/67948

The *sensory transfers* aim to recover, at least, protective sensation, avoiding skin ulcerations [52]. The sensory recovery, even if partial, will help with neuropathic pain control [12, 65].

End-to-end anastomosis: The distal stump of the donor nerve is coapted with the proximal stump of the recipient one. It is the best option for all nerve transfers, and the only successful one in motor restoration [66].

End-to-side anastomosis: The proximal end of an injured nerve is coapted to the side of a healthy one after creating an epineurial window in it [67]. The idea is that the axons of the healthy nerve create lateral sprouts that grow inside the damaged one. Although axons can travel in the epineurial space of rabbit nerves [18], in humans the epineurial window is essential to achieve axonal regeneration [68]. Sensory axons will spontaneously sprout from the healthy into the injured nerve, providing a kind of protective sensation, although it is never fully normal [11, 65, 66]. Meanwhile, donor nerve motor axons need an injury (crush or axotomy) to sprout inside an end-to-side coapted damaged nerve [66, 69]. It can be useful for sensory nerves but not for motor ones.

Reverse end-to-side or 'supercharge' end-to-side anastomosis: A healthy nerve is transected and coapted to the side of a damaged one [70]. The idea is that the axons of the healthy nerve grow inside the injured one. It provides a fast muscle re-innervation with preservation of the muscle bulk until the axons of the damaged nerve regenerate and reach their own endplates [66]. This avoids muscle atrophy while the injured nerve axons regenerate [71]. It has been used successfully in the case of anterior interosseous nerve (AIN) to motor fascicle of the UN transfer [52].

Single nerve transfer is when only one transfer is performed. *Dual nerve transfer* relates to the use of two different transfers to achieve the same function (i.e. shoulder abduction, elbow flexion, etc.).

In a *direct nerve transfer* donor and recipient nerves are coapted directly with no graft interposed. A tension-free suture is essential for a successful recovery [30].

2. Upper limb nerve transfers

The main *goals of upper limb re-innervation*, in order of importance, are as follows: (1) elbow flexion, (2) shoulder abduction and external rotation, (3) scapular stabilization, (4) elbow extension, (5) hand function and (6) sensory recovery in critical hand areas.

In *tetraplegic patients*, the most common *nerve transfers* are: *teres minor* nerve (TMN) to TLH nerve to recover elbow extension, *supinator* nerves (SNs) to PIN for hand and finger extension, ECRB to *flexor pollicis longus* nerve (FPLN) to recover thumb and index finger flexion and *brachialis* nerve branch (BCN) to AIN for thumb, index and middle finger flexion.

2.1. Scapular nerve transfers

Long thoracic nerve (LTN) damage is associated with scapular winging. To correct it one of the two branches of the *thoracodorsal nerve* (*TDN*) can be *coapted with the LTN* [72] (**Figure 1**). The re-innervation of the *serratus anterior* muscle (SAM) improves shoulder function [73].

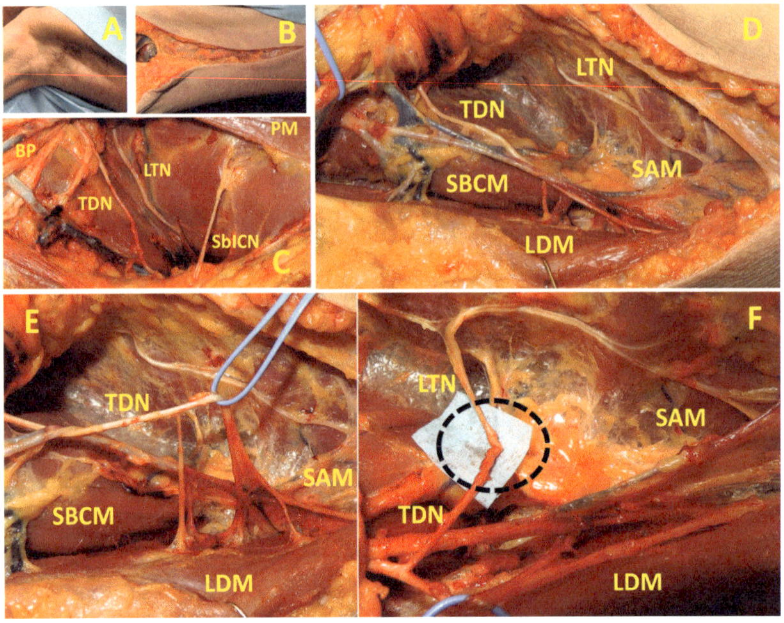

Figure 1. Thoracodorsal (TDN) to long thoracic nerve (LTN) transfer. One of the branches of the TDN is transferred to the LTN. Brachial plexus (BP), pectoralis minor muscle (PMM), sensory branch intercostal nerve (SbINC), subscapular muscle (SBCM), serratus anterior muscle (SAM), latissimus dorsi muscle (LDM).

2.2. Shoulder nerve transfers

Recovery of shoulder function is the second priority in brachial plexus injury. Shoulder abduction can be recovered with the *double nerve transfer SAN* branch for the trapezius to *SSN* together with *TLH nerve to the AXN* [45]. This dual shoulder transfers offer much better results than just one of them [17, 30].

In the *SAN to SSN transfer*, the distal branch of the SAN destined to the lower part of the trapezius muscle is transferred to the SSN, leaving the superior branches intact [38]. Results are much better with no nerve grafts interposed [17, 74]. As distal injury to the SSN can be present in addition to a more proximal brachial plexus damage, some recommend to dissect the SAN as distally as possible and to make the coaptation with the SSN as close as possible to the suprascapular notch [75, 76]. This transfer can be performed through an anterior or a posterior approach (**Figures 2** and **3**). In some, the posterior approach is better, as a smaller part of the trapezius muscle is denervated [77, 78]. Others disagree because it is technically difficult and because it cannot be done through a regular brachial plexus exploration [76], lengthening the surgical procedure [75]. The SAN should be used as a donor with caution if the *serratus anterior* muscle (SAM) is also paralysed for the risk of scapular winging [45].

First described by Lurje [79] in 1948, the *RAN to AXN transfer* was popularized by Leechavengvongs et al. [80] in 2003. There is no agreement on which is the best RAN branch to use [14, 38, 80]. Some surgeons report that, it is the TLH branch because it contains more motor axons, being the medial branch its alternative [81], but others think the opposite because

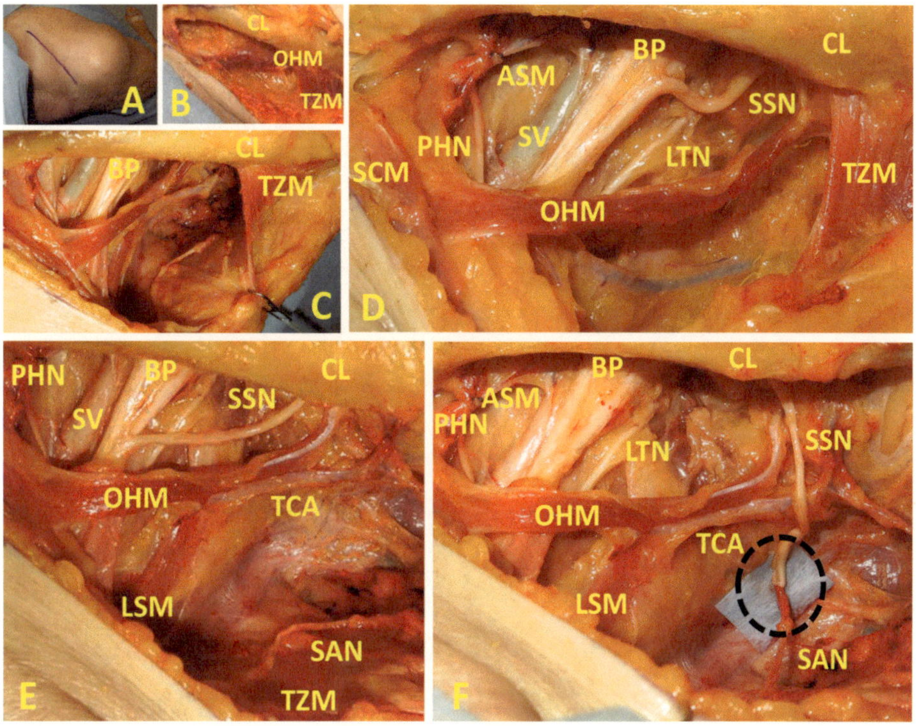

Figure 2. Spinal accessory (SAN) to suprascapular nerve (SSN) transfer (anterior approach). Sternocleidomastoid muscle (SCM), trapezius muscle (TZM), omohyoid muscle (OHM), brachial plexus (BP), anterior scalene muscle (ASM), clavicle (CL), phrenic nerve (PHN), transverse cervical artery (TCA), subclavian vein (SV), levator scapulae muscle (LSM).

as the TLH muscle inserts in the scapula it is part of the scapulohumeral joint [24] and because the medial head is longer and easier to isolate [82]. The results in shoulder abduction are good, particularly if combined with an SAN to SSN transfer [17, 38, 80]. As the recovery of shoulder external rotation is unsatisfactory [75, 83], many recommend to include the TMN in the transfer [38, 77, 80, 84]. The TLH to AXN transfer can be done through posterior to anterior approaches (**Figures 4** and **5**). The first one is the most widespread [85, 86], but the second is ideal if an Oberlin procedure is planned [87]. Both approaches show similar clinical outcomes [85, 87].

Other donor nerves that have been used to re-innervate the SSN and AXN are the C_3 and C_4 anterior rami [88], ICNs [89–92], TDN [60], MPN [93, 94], LTN [95], PHN [96], subscapular nerve [97], rhomboid nerve [98], ipsilateral or contralateral C_7 nerve root [99] and hypoglossal nerve [100]. They can be used but only if the SAN to SSN and TLH to AXN transfers are not possible, as their clinical outcome is unsatisfactory [17, 22].

2.3. Elbow flexion nerve transfers

This is the priority in brachial plexus injuries. First reported in 1994 [64], the *Oberlin procedure* is the transfer of the UN fascicle for the *flexor carpi ulnaris* (FCU) to the biceps muscle nerve (BMN) branch. To improve the results, some recommended a double nerve transfer adding the

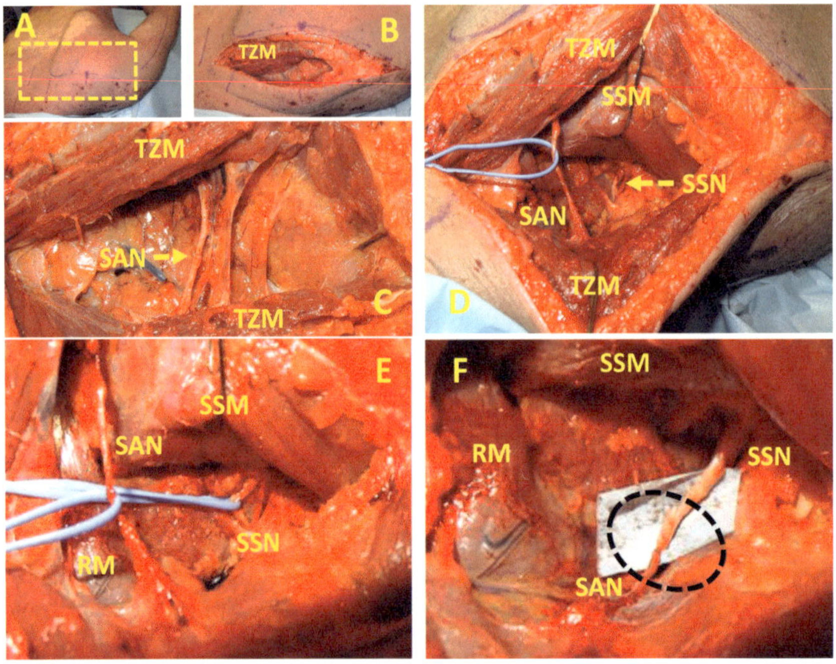

Figure 3. Spinal accessory (SAN) to suprascapular nerve (SSN) transfer (posterior approach). Trapezius muscle (TZM), supraspinatus muscle (SSM), rhomboid muscle (RM).

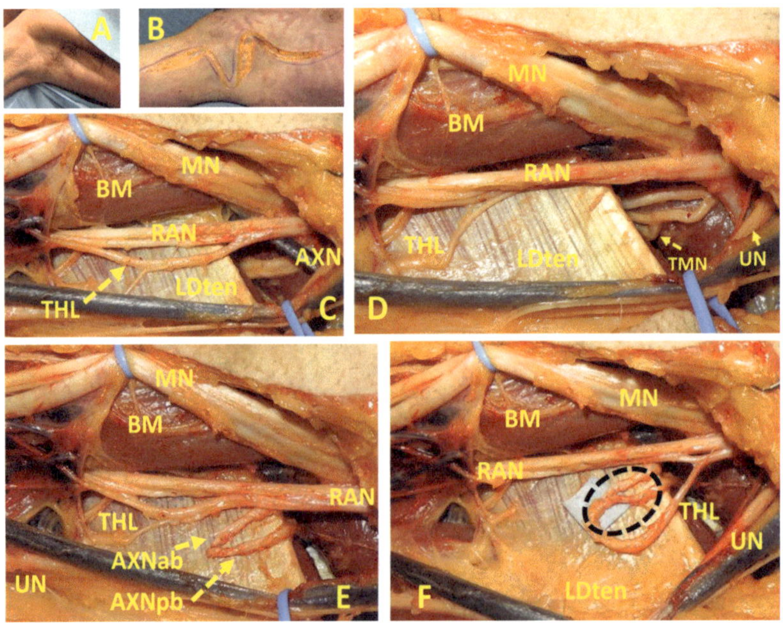

Figure 4. Radial nerve (RAN) to axillary nerve (AXN) transfer (anterior approach). Median nerve (MN), ulnar nerve (UN), triceps long head nerve branch (TLH), teres minor nerve branch (TMN), latissimus dorsi tendon (LDten), biceps muscle (BM), axillary nerve anterior branch (AXNab), axillary nerve posterior branch (AXNpb).

http://dx.doi.org/10.5772/67948

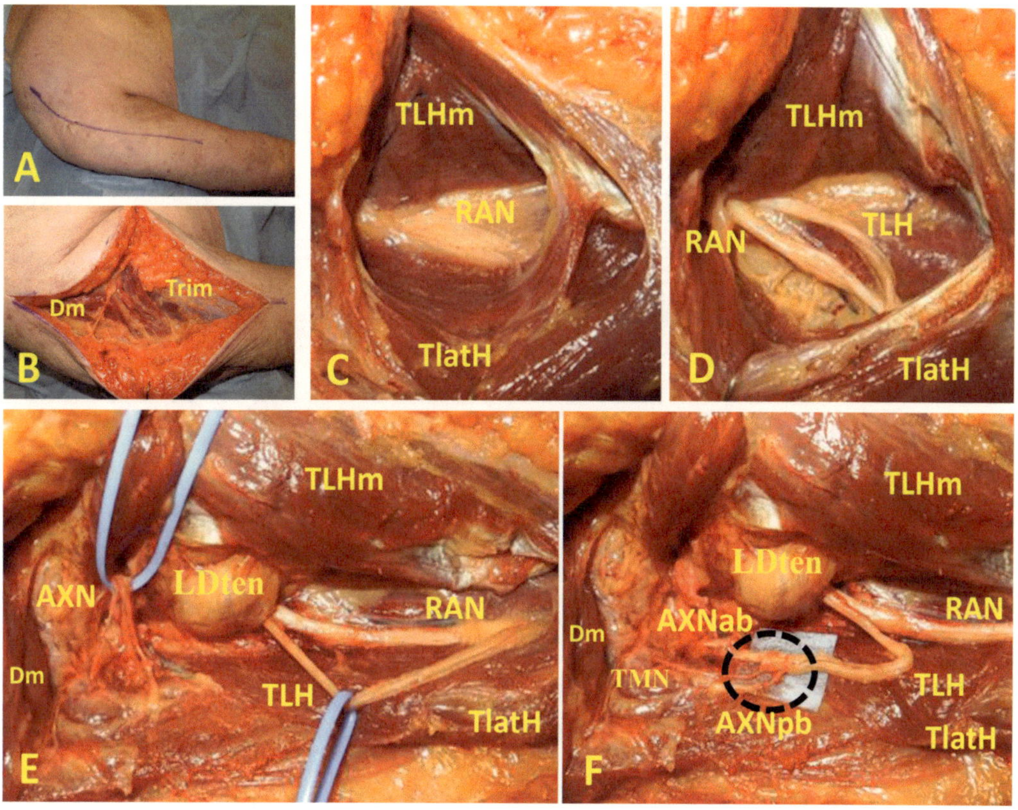

Figure 5. Radial nerve (RAN) to axillary nerve (AXN) transfer (posterior approach). Deltoid muscle (Dm), triceps muscle (Trim), triceps long head muscle (TLHm), triceps lateral head muscle (TlatH), triceps long head nerve branch (TLH), latissimus dorsi tendon (LDten), axillary nerve anterior branch (AXNab), axillary nerve posterior branch (AXNpb), teres minor nerve branch (TMN).

re-innervation of the BCM with the motor fascicle for the *flexor carpi radialis* (FCR) or the *flexor digitorum superficialis* (FDS), both from the MN [63, 101, 102] (**Figure 6**). This double transfer technique has lost popularity after two studies showed no difference in clinical outcome when compared with the Oberlin procedure [103, 104]. This procedure yields the best results provided the patient has a strong hand. Otherwise other alternatives must be considered.

ICN to MCN transfer: Two to three intercostal nerves should be transferred. The results are only fair, but can be the only choice in five root brachial plexus avulsion [105–107].

MPN to MCN transfer: Indicated in C_5–C_6 or C_5–C_6–C_7 nerve root injury with preservation of pectoralis muscle (PM) function [45]. This muscle is innervated by the superior, medial and lateral pectoral nerves, allowing it to retain some function when one of its branches is used as a donor [108] (**Figure 7**). It provides acceptable results [45, 63], but often a nerve graft is needed. This impairs the clinical outcome [109].

TDN to MCN transfer: Recommended if the Oberlin procedure is not possible [110, 111]. Not advised in the case of a weak shoulder adduction or if a muscle transfer is planned.

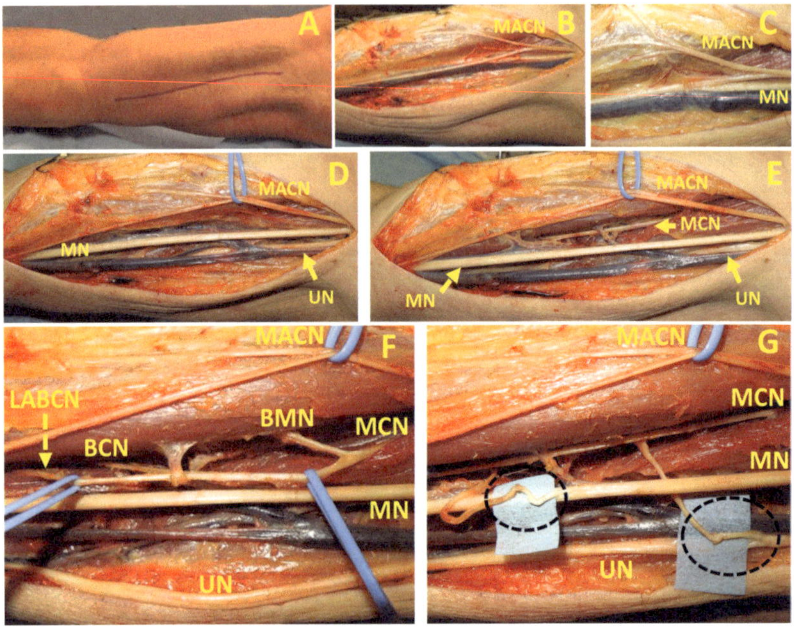

Figure 6. Dual nerve transfer ulnar nerve (UN) to biceps muscle nerve branch (BMN) and median nerve (MN) to brachialis muscle nerve (BCN) transfer. Medial antebrachial cutaneous nerve (MACN), musculocutaneous nerve (MCN), lateral antebrachial cutaneous nerve (LABCN).

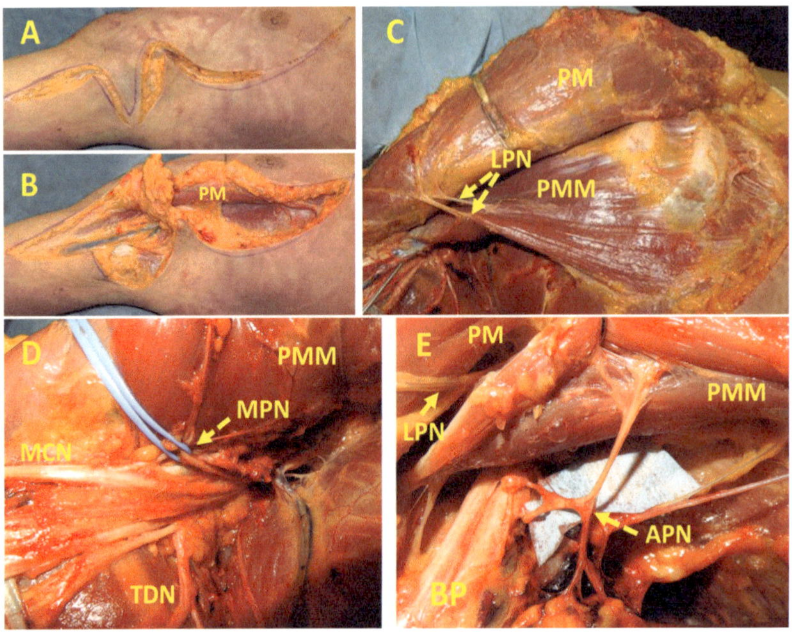

Figure 7. Medial pectoral nerve (MPN) to musculocutaneous nerve (MCN) transfer. Pectoralis major muscle (PM), pectoralis minor muscle (PMM), lateral pectoral nerve (LPN), thoracodorsal nerve (TDN), brachial plexus (BP), ansa pectoralis nerves (APN).

Other options: SAN to MCN [112, 113] and PHN to MCN [114]. Both usually need a nerve graft. None of them yield such good results as to recommend it [10].

2.4. Elbow extension nerve transfers

Although aided by gravity, there are many daily life activities that require active elbow extension (reaching overhead objects, changing from sitting to standing position, working over a table, throwing objects, changing from chair to bed in SCI patients, etc.) [115]. Restoration of elbow extension is particularly important in tetraplegia. The recipient nerve is usually the nerve branch for the TLH. The best results have been obtained by transferring the TMN to the TLH [116, 117]. Other possibilities are to re-innervate the TLH with ICNs [24, 92, 118, 119], a UN fascicle [24], the MPN [120], TDN [111], PHN [121], contralateral C_7 nerve root [122] and an RAN fascicle for the hand [123]. The results have been poor, particularly the ICNs [119].

2.5. Intercostal nerve transfers

First used by Seddon [16] in 1963 in brachial plexus repair. Only recommended when there is no other choice (i.e. C_5–T_1 brachial plexus avulsion). Harvesting them is technically demanding, requiring arterial hypotension to control the bleeding as the cautery cannot be used until the ICN is fully harvested [24]. Up to seven ICNs can be transferred. The first one was used by Durand et al. [91] in a single case. Harvesting the second one is not advised as it provides a large sensory contribution to the arm [24] (**Figure 8**). Additionally, it is technically very

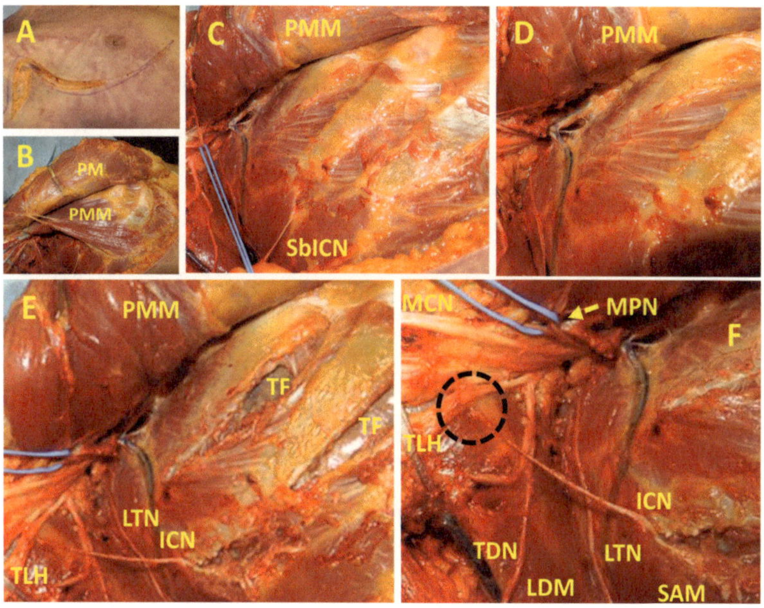

Figure 8. Intercostal nerve (ICN) to triceps long head nerve branch (TLH) of radial nerve (RA). Pectoralis major muscle (PM), pectoralis minor muscle (PMM), lateral pectoral nerve (LPN), thoracodorsal nerve (TDN), brachial plexus (BP), sensory branch intercostal nerve (SbICN), thoracodorsal nerve (TDN), long thoracic nerve (LTN), medial pectoral nerve (MPN), latissimus dorsi muscle (LDM), serratus anterior muscle (SAM), thoracic fascia (TF).

difficult to harvest and a nerve graft is always needed [24]. Usually, the third to the fifth intercostal nerves are the ones used. In women, the fourth one should be preserved to retain the nipple's area sensation. After harvesting the mean available, ICN length is 11–12 cm [124], so no nerve graft is usually needed. At least two of them per recipient nerve are needed [24, 124]. They have been used to re-innervate many nerves, like the AXN [89, 92], MCN [24, 105, 106, 124–126], TLH [92, 118, 119], TDN [89, 127] and SSN [90, 91]. Their sensory branches can be used to recover some limb sensation, ameliorating the neuropathic pain. Unfortunately, not being synergistic with the recipients nerves a long re-education must be expected [118]. Usually their harvest is not associated with any pulmonary dysfunction [24, 128].

2.6. Phrenic nerve transfers

It has been used in C_5–T_1 nerve root brachial plexus avulsion to re-innervate the MCN [106], the median nerve (MN) [129] or the RAN [46]. Unless its whole intra-thoracic segment is harvested, a nerve graft is needed [129]. The clinical results are acceptable provided there is no other choice. Patients usually need to take a breath before starting the movement with the re-innervated muscle. Post-operatively patients show a decreased pulmonary capacity that improves after 2 years [114, 130]. It is not commonly used these days.

2.7. Contralateral C_7 nerve transfer

It has been used in complete brachial plexus avulsions. It requires a long nerve graft or to shorten the humerus [122, 131], but this can be avoided crossing the donor C_7 nerve root through the C_6–C_7 disc space [132]. The targets are usually MN or MCN. Its use has been discouraged as its clinical results are poor and unreliable, the forearm and hand muscles do not get a good re-innervation and there is co-contraction of the donor and recipient limbs [24, 43]. In fact, initiating the movement often requires to start it in the contralateral normal side and there is co-contraction of the muscles of the donor and recipient sides [24]. Its value is very much disputed.

2.8. Wrist and finger extension

In the case of proximal RAN damage with an intact MN, the best choice is the *dual nerve transfer* described by Mackinnon et al. in 2007 [133]. The FDS branches (usually there are two) are coapted to the ECRB branch to recover wrist extension [134] and *the branches for the FCR and palmaris longus* (*PL*) are coapted *with the PIN* to recover finger extension [32, 50, 135] (**Figure 9**). This combination provides the best outcome as muscles are synergistic [29]. Contrariwise to tendon transfers, this dual nerve transfer recovers independent finger extension [11, 32, 135].

Bertelli and Ghizoni reported successful restoration of wrist extension transferring the *PQ motor branch to the ECRB* [136] (**Figure 10**). This technique is very useful but is more technically demanding than the previous one.

The pronator *teres* (PT) branch is not recommended as a donor for RAN re-innervation because pronation is essential for many daily living activities and because this muscle can be used for a tendon transfer in case the nerve transfer is unsuccessful [31, 135].

http://dx.doi.org/10.5772/67948

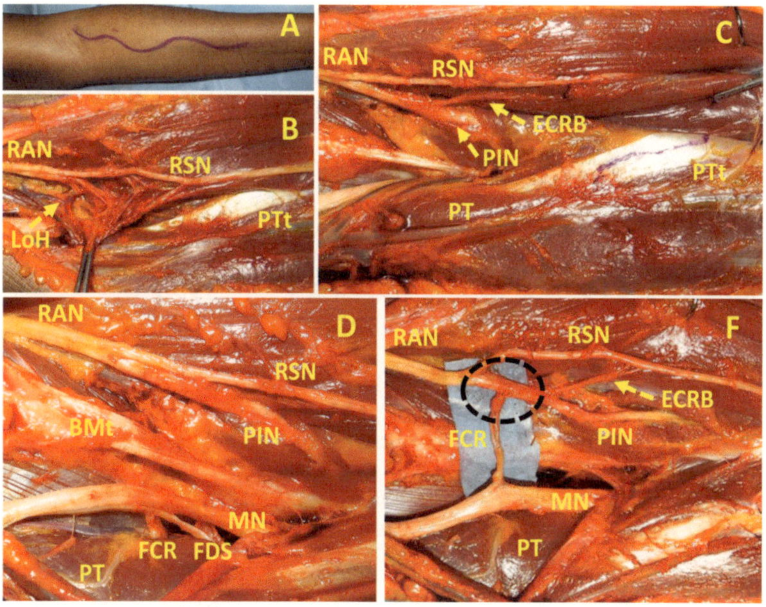

Figure 9. Flexor digitorum superficialis nerve (FDS) to extensor carpi radialis brevis (ECRB) nerve transfer and flexor carpi radialis (FCR) to posterior interosseous nerve (PIN) transfer. Radial nerve (RAN), median nerve (MN), radial sensory nerve (RSN), pronator teres muscle (PT), pronator teres tendon (PTt), lash of henry (LoH), biceps muscle tendon (BMt).

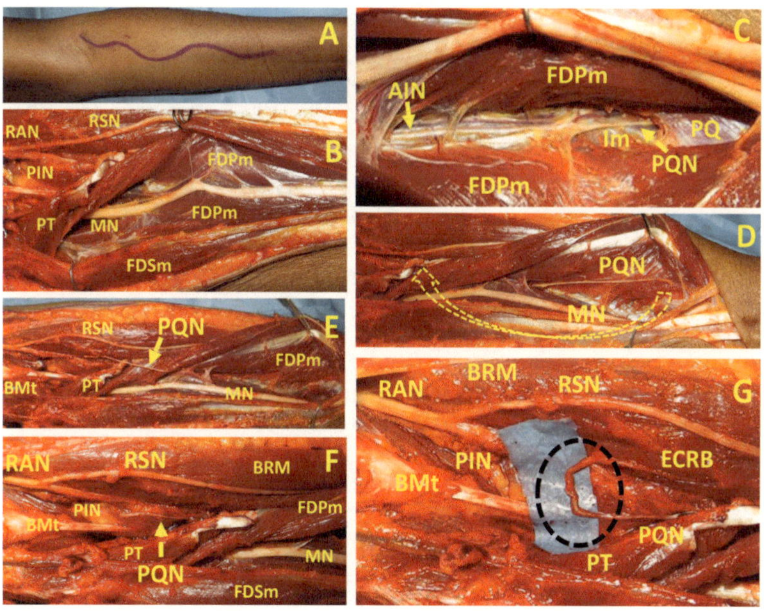

Figure 10. Pronator quadratus nerve branch (PQN) from anterior interosseous nerve (AIN) to extensor carpi radialis brevis nerve (ECRB) transfer. Pronator quadratus muscle (PQ), radial nerve (RAN), radial sensory nerve (RSN), median nerve (MN), pronator teres muscle (PT), flexor digitorum profundus muscle (FDPm), flexor digitorum superficialis muscle (FDSm), brachioradialis muscle (BRM), interosseous membrane (Im).

In C_7–T_1 brachial plexus injuries, the shoulder and elbow mobility is preserved, but the finger movements are lost. The supinator muscle (SM) innervation is preserved as is comes from the C_6 nerve root. Thus, it is possible to coapt one or both *SN branches to the PIN* (**Figure 11**). This helps to recover the function of the *extensor pollicis longus* (EPL) and *extensor digitorum communis* (EDC) muscles. It is a very successful nerve transfer [137–139].

2.9. Finger flexion and median nerve hand function

The primary goal of MN recovery is to provide first and second finger pincer as well as thumb opposition [29].

Reported by Palazzi et al. [140] in 2006 the *BCN to AIN nerve transfer* is indicated in the case of C_8–T_1 nerve root or lower brachial plexus injury with a MN dysfunction. It provides improvement in hand function [13, 141–143] but there are other transfers with better clinical results. It is recommended only if the MCN nerve is intact.

García-López et al. [144] reported a single case of *brachioradialis muscle* (*BRM*) *nerve to AIN nerve* transfer with return of finger flexion.

Thenar muscle re-innervation can be achieved coapting the *AIN terminal branch to the PQ with the MN motor thenar branch* [145]. A nerve graft is usually required [29] but it is a very effective procedure [29, 145, 146].

The *FDS, FCR* and *FCU branches* can all be transferred *to recover AIN function* [147].

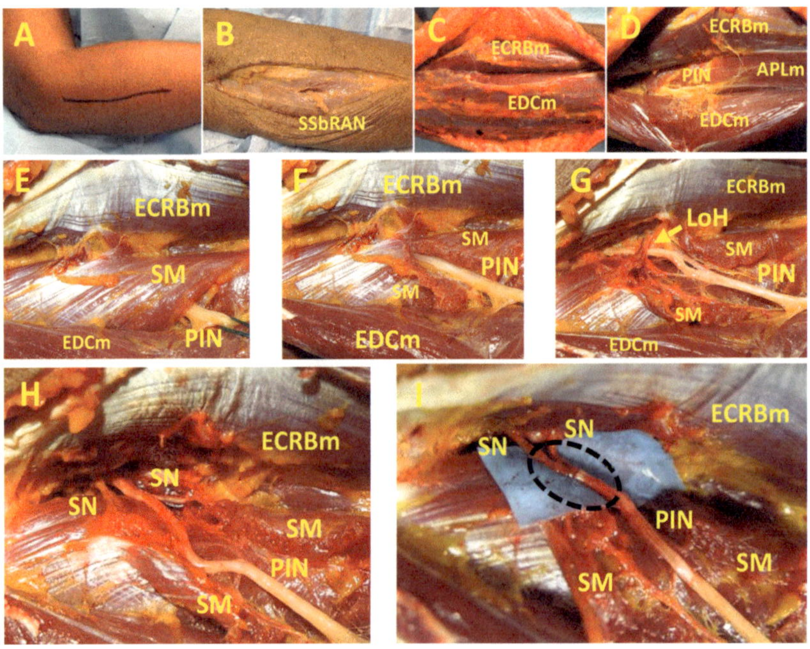

Figure 11. Supinator nerve branches (SN) to posterior interosseous nerve (PIN) transfer. Superficial sensory branch radial nerve (SSbRAN), supinator muscle (SM), extensor carpi radialis brevis muscle (ECRBm), extensor digitorum communis muscle (EDCm), lash of Henry (LoH), abductor pollicis longus muscle (APLm).

Finger flexion, particularly the thumb and index finger, can be recovered by coapting the *nerve branch for the ECRB to the AIN* [29]. *One or both SM branches can be coapted with the AIN* to recover FPL and FDP function [142]. Some have transferred both the ECRB and SM branches to the AIN to recover finger flexion [148]. The results of these techniques are very encouraging [149].

2.10. Pronation recovery

To restore active pronation, essential in many daily living activities, some have transferred the FCU nerve branch to the PT nerve [150]. Others have transferred one of the branches of the FDS to the PT nerve [40]. More recently, the nerve for the ECRB has been transferred to the PT branch [142]. This last technique has a widespread acceptance.

2.11. Ulnar nerve hand function

When the UN is damaged in the arm or more proximally, the recovery of hand intrinsic muscles is dismal [151]. The motor recovery can be improved by coapting the *AIN terminal branch for the PQ to the motor fascicle of the UN* [50, 62, 146, 152, 153] (**Figure 12**). This technique was originally described in 1997 by Wang and Zhu [146]. If this fascicle is neurolised proximally enough in the forearm, no nerve graft is needed [29]. The clinical outcome is very satisfactory [29, 62, 152]. Another possibility is a reverse end-to-side transfer, avoiding atrophy of the UN innervated hand muscles while awaiting for the native UN motor axons to regenerate [70, 71, 154]. It is very effective [29].

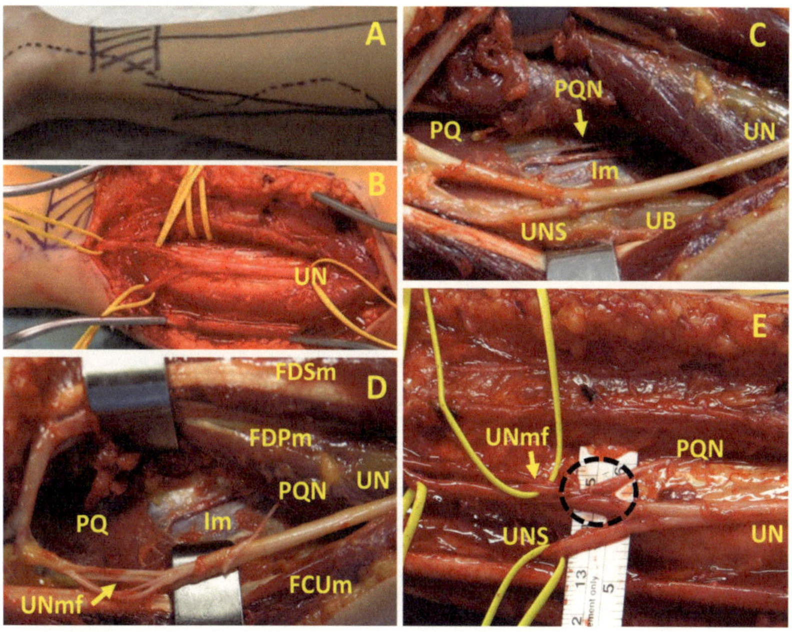

Figure 12. Pronator quadratus nerve branch (PQN) from anterior interosseous nerve (AIN) to ulnar nerve (UN) motor fascicle (UNmf) nerve transfer. Pronator quadratus muscle (PQ), interosseous membrane (Im), ulnar nerve dorsal sensory branch (UNDS), ulnar bone (UB), flexor carpi ulnaris muscle (FCUm), flexor digitorum superficialis muscle (FDSm), flexor digitorum profundus muscle (FDPm).

Transfer of the *terminal branches of the extensor digiti minimi* (*EDM*) and *the extensor carpi ulnaris* (*ECU*) *to the UN motor branch* in the forearm has been reported. In this single case, a 10-cm nerve graft had to be used and the recovery was fair [155]. It is not recommended unless there is no other option.

2.12. Nerve transfers for upper limb sensation recovery

Sensory nerve transfers sacrifice nerves that serve areas with non-critical sensation to recover it where this sense is vital (i.e. tip of the thumb and index fingers) [33, 49, 51, 61, 65]. They also help with neuropathic pain control [13, 29, 49, 156]. The donor nerve distal stump can be coapted end-to-side to a nearby sensory nerve to regain some protective sensation [14, 31, 66].

In the case of proximal MN damage, restoring the sensation in the ulnar aspect of the thumb and the radial side of the index finger is a priority to allow a useful pincer mechanism [29]. The donors are the nerves supplying less essential areas like the third web space (MN branch) the forth web space (UN branch), the dorsal sensory branch of the UN and the radial sensory nerve (RSN) [22, 29, 49, 50, 65, 157, 158]. Ducic et al. [159] in 2006 recovered the sensation of the thumb and index fingers using the radial nerve branches as donors. Bertelli and Ghizoni [158] in 2011 reported the transfer of the distal superficial radial finger nerves of the thumb and index fingers to the ulnar aspect of the thumb and the radial aspect of the index finger. Flores et al. [52] transferred the superficial ulnar nerve to the third palmar digital nerve.

In the case of proximal RAN damage, the sensation in the dorsum of the hand can be recovered by coapting the lateral antebrachial cutaneous nerve (LABCN) to the RSN [148] (**Figure 13**). At

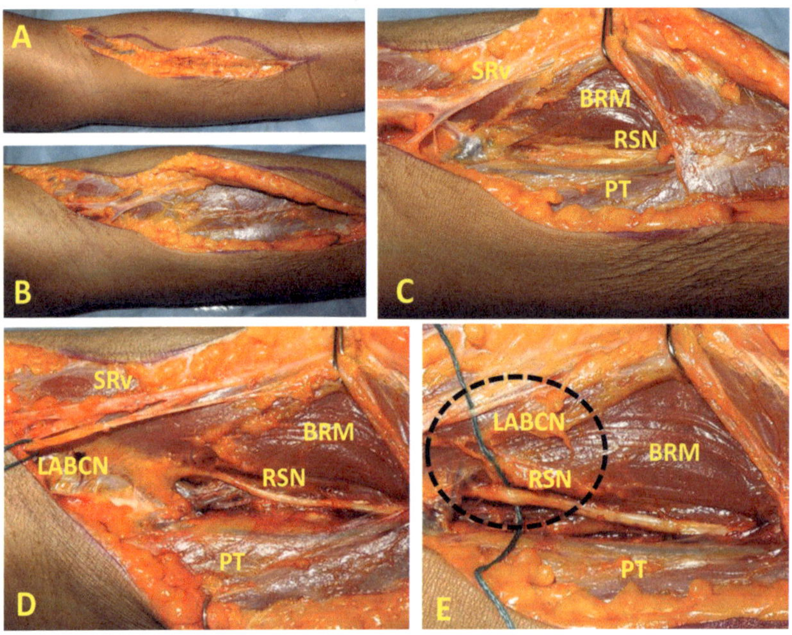

Figure 13. Lateral antebrachial cutaneous nerve (LABCN) to radial sensory nerve (RSN) transfer. Superficial radial vein (SRv), brachioradialis muscle (BRM), pronator teres muscle (PT).

the outer aspect of the elbow, both nerves run parallel to each other. The LABCN goes with the superficial radial vein and the RSN with the deep radial vessels. The LABCN is sectioned as distally as possible and coapted to the proximal part of the RSN [17].

Protective sensation in the hand ulnar innervated areas can be recovered with a sensory nerve transfer from the third web space (MN branch) [29] or from the RSN [160].

In C_7–T_1 nerve root injuries the LABCN can be coapted with the sensory fascicles of the UN in the forearm to recover the sensation in the hand [12].

3. Nerve transfers in the lower extremity

Three areas have been explored. At the proximal level, nerve transfers between the femoral (FN) and obturator (OBN) nerves; at the knee, the nerve transfers between branches of the posterior tibial nerve (PTN) and the peroneal nerve (PN); and at the foot, some sensory nerve transfers for recovery of protective sensation. There have been some attempts to recover urinary continence in spinal injured patients.

FN injuries are the most disabling as they impair the capacity of standing and walking. The anterior branch of the OBN has been successful in re-innervating the FN [119, 161, 162]. The anterior branch of the OBN is selected to preserve the primary leg adductor muscle [162]. The reverse transfer using an FN motor branch as donor to coapt it to the OBN has also been used with successful clinical outcomes [163].

PTN to PN transfer to recover foot dorsiflexion has been attempted [164–166] but the reported outcomes are inferior to the posterior tibial tendon transfer [167].

Nerve transfer from an FN branch to the pudendal nerve to recover urinary continence has been achieved in a canine neurogenic bladder model [168] but its clinical application in the human being is still pending.

4. Transfers for spinal cord injured patients

Around a 50% of the SCI involves the cervical spinal cord [30]. Tetraplegic patients suffer from a variable loss of arm and hand functions, usually asymmetrical [30]. This creates serious difficulties to perform their daily activities (transfers back and forth from the wheelchair, feeding, computer handling, self-catheterization, etc.) [169]. To recover any partial arm and/or hand function might mean an immense impact on their quality of life [170].

Nerve transfers used in SCI patients aim to recover elbow, wrist and finger extension, palmar grasp and thumb and index finger pinch and release. The nerve transfers mentioned above can also be used in these cases. The only difference is that in SCI there are three spinal cord areas [170–172]. The area above the injury will have normal spinal cord and nerves; the area of the injury will have neuronal loss and severe muscle atrophy. To recover function in this area, nerve transfers must be planned no later than 1 year after injury.

Below is the area of normal spinal cord disconnected from the rest of the central nervous system. The muscles depending from this area are not denervated, so nerve transfers can be done any time. The first area is the donor and second and third areas are the recipients. We aim to recover some functions of areas 2 and 3 transferring some nerve branches from area 1. In SCI, it is recommended to wait 12 months to give a chance for spontaneous recovery [172].

Elbow extension has been achieved by transferring the TM motor branch to the TLH [116].

Wrist and finger extension can be recovered with SN to PIN transfer [173, 174].

Thumb and index finger pinch is restored transferring the ECRB motor branch to the AIN [175, 176].

The BCM branch has been transferred to the AIN fascicle in the arm to recover FPL and FDP function [147, 177, 178].

5. Conclusions

Nerve transfers have become an essential way to repair irrecoverable peripheral nerve injuries. Mastering these techniques is essential for the peripheral nerve surgeon. The best results are obtained when patients are young; the procedure is not delayed more than 6 months after the injury, donor and recipient nerves are agonistic and when the donor nerve has no damage. Shoulder and elbow motor recovery is very successful in most patients. Hand recovery can also be achieved but results are not so good. The biggest experience is in upper limb nerve transfers. Lower limb nerve transfers have been attempted but the experience is limited and the choices few. In tetraplegic patients, we aim to recover some of the lost functions, simplifying their daily lives.

Acknowledgements

The authors thank the Department of Human Anatomy and Embryology of the Faculty of Medicine of the University of Valencia, particularly the laboratory curators Lucia and Carmina and Dr. Tomás Hernández Gil de Tejada and all personnel of the *Instituto de Medicina Legal de Valencia* for their assistance in this study.

Appendices and nomenclatures

AIN	Anterior interosseous nerve.
AXN	Axillary nerve.
BM	Biceps muscle.
BMN	Biceps muscle nerve.

http://dx.doi.org/10.5772/67948

BCM	Brachialis muscle.
BCN	Brachialis nerve
BRM	Brachioradialis muscle.
BRN	Brachioradialis nerve.
ECRB	Extensor carpi radialis brevis.
ECU	Extensor carpi ulnaris.
EDC	Extensor digitorum communis.
EDM	Extensor digiti minimi.
EPL	Extensor pollicis longus.
FCR	Flexor carpi radialis.
FCU	Flexor carpi ulnaris.
FDS	Flexor digitorum superficialis.
FDP	Flexor digitorum profundus.
FN	Femoral nerve.
FPLN	Flexor pollicis longus.
ICN	Intercostal nerve.
LABCN	Lateral antebrachial cutaneous nerve.
LTN	Long thoracic nerve.
MCN	Musculocutaneous nerve.
MN	Median nerve.
MPN	Medial pectoral nerve.
MRC	Medical Research Council.
OBN	Obturator nerve.
PHN	Phrenic nerve.
PIN	Posterior interosseous nerve.
PL	Palmaris longus.
PM	Pectoralis muscle.
PN	Peroneal nerve.
PQ	Pronator quadratus.
PT	Pronator teres.
PTN	Posterior tibial nerve.
RAN	Radial nerve.
RSN	Radial sensory nerve.
SAM	Serratus anterior muscle.
SAN	Spinal accessory nerve.

SAM	Serratus anterior muscle.
SCI	Spinal cord injury.
SM	Supinator muscle.
SN	Supinator nerve.
SSN	Suprascapular nerve.
TDN	Thoracodorsal nerve.
TLH	Triceps long head.
TMN	Teres minor nerve.
UN	Ulnar nerve.

Author details

Vicente Vanaclocha-Vanaclocha[1]*, Jose María Ortiz-Criado[2], Nieves Sáiz-Sapena[3] and Nieves Vanaclocha[1]

*Address all correspondence to: vvanaclo@hotmail.com

1 Department of Neurosurgery, Hospital General Universitario, Valencia, Spain

2 Universidad Católica San Vicente Mártir, Valencia, Spain

3 Hospital 9 de Octubre, Valencia, Spain

References

[1] Bentolila V, Nizard R, Bizot P, Sedel L. Complete traumatic brachial plexus palsy. Treatment and outcome after repair. J Bone Joint Surg Am. 1999 Jan;81(1):20–8.

[2] Seddon HJ, Medawar PB, Smith H. Rate of regeneration of peripheral nerves in man. J Physiol. 1943 Sep 30;102(2):191–215.

[3] Isaacs J. Treatment of acute peripheral nerve injuries: current concepts. J Hand Surg. 2010 Mar;35(3):491–497; quiz 498.

[4] Dahlin LB. The role of timing in nerve reconstruction. Int Rev Neurobiol. 2013;109:151–64.

[5] Finkelstein DI, Dooley PC, Luff AR. Recovery of muscle after different periods of denervation and treatments. Muscle Nerve. 1993 Jul;16(7):769–77.

[6] Fu SY, Gordon T. Contributing factors to poor functional recovery after delayed nerve repair: prolonged denervation. J Neurosci. 1995 May;15(5):3886–95.

[7] Kobayashi J, Mackinnon SE, Watanabe O, Ball DJ, Gu XM, Hunter DA, Kuzon WM. The effect of duration of muscle denervation on functional recovery in the rat model. Muscle Nerve. 1997 Jul;20(7):858–66.

[8] Aydin MA, Mackinnon SE, Gu XM, Kobayashi J, Kuzon WM. Force deficits in skeletal muscle after delayed reinnervation. Plast Reconstr Surg. 2004 May;113(6):1712–8.

[9] Bertelli JA, Ghizoni MF. Nerve root grafting and distal nerve transfers for C5-C6 brachial plexus injuries. J Hand Surg. 2010 May;35(5):769–75.

[10] Ray WZ, Chang J, Hawasli A, Wilson TJ, Yang L. Motor nerve transfers: a comprehensive review. Neurosurgery. 2016 Jan;78(1):1–26.

[11] Boyd KU, Nimigan AS, Mackinnon SE. Nerve reconstruction in the hand and upper extremity. Clin Plast Surg. 2011 Oct;38(4):643–60.

[12] Lee SK, Wolfe SW. Nerve transfers for the upper extremity: new horizons in nerve reconstruction. J Am Acad Orthop Surg. 2012 Aug;20(8):506–17.

[13] Tung TH, Mackinnon SE. Nerve transfers: indications, techniques, and outcomes. J Hand Surg. 2010 Feb;35(2):332–41.

[14] Mackinnon SE, Colbert SH. Nerve transfers in the hand and upper extremity surgery. Tech Hand Up Extrem Surg. 2008 Mar;12(1):20–33.

[15] Terzis J, Faibisoff B, Williams B. The nerve gap: suture under tension vs. graft. Plast Reconstr Surg. 1975 Aug;56(2):166–70.

[16] Seddon HJ. Nerve grafting. J Bone Joint Surg Br. 1963 Aug;45:447–61.

[17] Merrell GA, Barrie KA, Katz DL, Wolfe SW. Results of nerve transfer techniques for restoration of shoulder and elbow function in the context of a meta-analysis of the English literature. J Hand Surg. 2001 Mar;26(2):303–14.

[18] McCallister WV, McCallister EL, Trumble SA, Trumble TE. Overcoming peripheral-nerve gap defects using an intact nerve bridge in a rabbit model. J Reconstr Microsurg. 2005 Apr;21(3):197–206.

[19] Dvali L, Mackinnon S. Nerve repair, grafting, and nerve transfers. Clin Plast Surg. 2003 Apr;30(2):203–21.

[20] Poppler LH, Davidge K, Lu JCY, Armstrong J, Fox IK, Mackinnon SE. Alternatives to sural nerve grafts in the upper extremity. Hand (NY). 2015 Mar;10(1):68–75.

[21] Nichols CM, Brenner MJ, Fox IK, Tung TH, Hunter DA, Rickman SR, Mackinnon SE. Effects of motor versus sensory nerve grafts on peripheral nerve regeneration. Exp Neurol. 2004 Dec;190(2):347–55.

[22] Moore AM. Nerve transfers to restore upper extremity function: a paradigm shift. Front Neurol. 2014;5:40.

[23] Gordon T, Eva P, Borschel GH. Delayed peripheral nerve repair: methods, including surgical "cross-bridging" to promote nerve regeneration. Neural Regen Res. 2015 Oct;10(10):1540–4.

[24] Oberlin C, Durand S, Belheyar Z, Shafi M, David E, Asfazadourian H. Nerve transfers in brachial plexus palsies. Chir Main. 2009 Feb;28(1):1–9.

[25] Chan KM, Olson JL, Morhart M, Lin T, Guilfoyle R. Outcomes of nerve transfer versus nerve graft in ulnar nerve laceration. Can J Neurol Sci. 2012 Mar;39(2):242–4.

[26] Haftek J. Autogenous cable nerve grafting instead of end to end anastomosis in secondary nerve suture. Acta Neurochir (Wien). 1976;34(1–4):217–21.

[27] Mackinnon SE. Donor distal, recipient proximal and other personal perspectives on nerve transfers. Hand Clin. 2016 May;32(2):141–51.

[28] Ropars M, Dréano T, Siret P, Belot N, Langlais F. Long-term results of tendon transfers in radial and posterior interosseous nerve paralysis. J Hand Surg Edinb Scotl. 2006 Oct;31(5):502–6.

[29] Brown JM, Mackinnon SE. Nerve transfers in the forearm and hand. Hand Clin. 2008 Nov;24(4):319–340 v.

[30] Garg R, Merrell GA, Hillstrom HJ, Wolfe SW. Comparison of nerve transfers and nerve grafting for traumatic upper plexus palsy: a systematic review and analysis. J Bone Joint Surg Am. 2011 May 4;93(9):819–29.

[31] Moore AM, Franco M, Tung TH. Motor and sensory nerve transfers in the forearm and hand. Plast Reconstr Surg. 2014 Oct;134(4):721–30.

[32] Ray WZ, Mackinnon SE. Clinical outcomes following median to radial nerve transfers. J Hand Surg. 2011 Feb;36(2):201–8.

[33] Rinker B. Nerve transfers in the upper extremity: a practical user's guide. Ann Plast Surg. 2015 Jun;74(Suppl 4):S222–8.

[34] Gordon T, Yang JF, Ayer K, Stein RB, Tyreman N. Recovery potential of muscle after partial denervation: a comparison between rats and humans. Brain Res Bull. 1993; 30(3–4):477–82.

[35] Lien SC, Cederna PS, Kuzon WM. Optimizing skeletal muscle reinnervation with nerve transfer. Hand Clin. 2008 Nov;24(4):445–54, vii.

[36] Rafuse VF, Pattullo MC, Gordon T. Innervation ratio and motor unit force in large muscles: a study of chronically stimulated cat medial gastrocnemius. J Physiol. 1997 Mar 15;499(Pt 3):809–23.

[37] Gordon T, Tyreman N, Raji MA. The basis for diminished functional recovery after delayed peripheral nerve repair. J Neurosci. 2011, 31 (14) 5325–34.

[38] Bertelli JA, Ghizoni MF. Reconstruction of C5 and C6 brachial plexus avulsion injury by multiple nerve transfers: spinal accessory to suprascapular, ulnar fascicles to biceps branch, and triceps long or lateral head branch to axillary nerve. J Hand Surg. 2004 Jan;29(1):131–9.

[39] Weber RV, Mackinnon SE. Bridging the neural gap. Clin Plast Surg. 2005 Oct;32(4):605–616, viii.

[40] Tung TH, Mackinnon SE. Flexor digitorum superficialis nerve transfer to restore pronation: two case reports and anatomic study. J Hand Surg. 2001 Nov;26(6):1065–72.

[41] Yang J, Jia X, Yu C, Gu Y. Pronator teres branch transfer to the anterior interosseous nerve for treating C8T1 brachial plexus avulsion: an anatomic study and case report. Neurosurgery. 2014 Oct;75(4):375–9; discussion 379.

[42] Wong AH, Pianta TJ, Mastella DJ. Nerve transfers. Hand Clin. 2012 Nov;28(4):571–7.

[43] Sammer DM, Kircher MF, Bishop AT, Spinner RJ, Shin AY. Hemi-contralateral C7 transfer in traumatic brachial plexus injuries: outcomes and complications. J Bone Joint Surg Am. 2012 Jan 18;94(2):131–7.

[44] Liu Y, Xu X-C, Zou Y, Li S-R, Zhang B, Wang Y. Phrenic nerve transfer to the musculocutaneous nerve for the repair of brachial plexus injury: electrophysiological characteristics. Neural Regen Res. 2015 Feb;10(2):328–33.

[45] Hems T. Nerve transfers for traumatic brachial plexus injury: advantages and problems. J Hand Microsurg. 2011 Jun;3(1):6–10.

[46] Wang S, Li P, Xue Y, Zou J, Li W, Li Y. Direct coaptation of the phrenic nerve with the posterior division of the lower trunk to restore finger and elbow extension function in patients with total brachial plexus injuries. Neurosurgery. 2016 Feb;78(2):208–15.

[47] Messina A, Van Zyl N, Weymouth M, Flood S, Nunn A, Cooper C, Hahn J, Galea MP. Morphology of donor and recipient nerves utilised in nerve transfers to restore upper limb function in cervical spinal cord injury. Brain Sci. 2016 Sep 27;6, 42; doi:10.3390/brainsci6040042.

[48] Anastakis DJ, Chen R, Davis KD, Mikulis D. Cortical plasticity following upper extremity injury and reconstruction. Clin Plast Surg. 2005 Oct;32(4):617–34, viii.

[49] Brunelli GA. Sensory nerves transfers. J Hand Surg Edinb Scotl. 2004 Dec;29(6):557–62.

[50] Brown JM, Yee A, Mackinnon SE. Distal median to ulnar nerve transfers to restore ulnar motor and sensory function within the hand: technical nuances. Neurosurgery. 2009 Nov;65(5):966–78.

[51] Suppaphol S, Watcharananan I, Tawonsawatruk T, Woratanarat P, Sasivongsbhakdi T, Kawinwonggowit V. The sensory restoration in radial nerve injury using the first branch of dorsal ulnar cutaneous nerve—a cadaveric study for the feasibility of procedure and case demonstration. J Med Assoc Thail Chotmaihet Thangphaet. 2014 Mar;97(3):328–32.

[52] Flores LP. Distal anterior interosseous nerve transfer to the deep ulnar nerve and end-to-side suture of the superficial ulnar nerve to the third common palmar digital nerve for treatment of high ulnar nerve injuries: experience in five cases. Arq Neuropsiquiatr. 2011 Jun;69(3):519–24.

[53] Ballance CA, Ballance HA, Stewart P. Remarks on the operative treatment of chronic facial palsy of peripheral origin. Br Med J. 1903 May 2;1(2209):1009–13.

[54] Tuttle H. Exposure of the brachial plexus with nerve transplantation. JAMA. 1913;61:15–7.

[55] Vulpius, O, Stoffel, A. Orthopädische Operationslehre. 2nd Edition. Stuttgart: Enke; 1920.

[56] Meals RA, Nelissen RG. The origin and meaning of "neurotization." J Hand Surg. 1995 Jan;20(1):144–6.

[57] Harris RI. The treatment of irreparable nerve injuries. Can Med Assoc J. 1921 Nov;11(11):833–41.

[58] Förster, O. Die Therapie Des Shussverletzungen Der Peripheren Nerven. In: Handbuch der Neurologie (Lewandowsky). Berlin: Ergänzungsband; 1929. p. chapter 3.

[59] Lurje AS. On the use of N. Musculocutaneous for neurotization of N. Radialis in cases of very large defects of the latter. Ann Surg. 1948 Jul;128(1):110–5.

[60] Samardzic MM, Grujicic DM, Rasulic LG, Milicic BR. The use of thoracodorsal nerve transfer in restoration of irreparable C5 and C6 spinal nerve lesions. Br J Plast Surg. 2005 Jun;58(4):541–6.

[61] Bedeschi P, Celli L, Balli A. Transfer of sensory nerves in hand surgery. J Hand Surg Edinb Scotl. 1984 Feb;9(1):46–9.

[62] Novak CB, Mackinnon SE. Distal anterior interosseous nerve transfer to the deep motor branch of the ulnar nerve for reconstruction of high ulnar nerve injuries. J Reconstr Microsurg. 2002 Aug;18(6):459–64.

[63] Brandt KE, Mackinnon SE. A technique for maximizing biceps recovery in brachial plexus reconstruction. J Hand Surg. 1993 Jul;18(4):726–33.

[64] Oberlin C, Béal D, Leechavengvongs S, Salon A, Dauge MC, Sarcy JJ. Nerve transfer to biceps muscle using a part of ulnar nerve for C5-C6 avulsion of the brachial plexus: anatomical study and report of four cases. J Hand Surg. 1994 Mar;19(2):232–7.

[65] Bertelli JA. Distal sensory nerve transfers in lower-type injuries of the brachial plexus. J Hand Surg. 2012 Jun;37(6):1194–9.

[66] Hayashi A, Pannucci C, Moradzadeh A, Kawamura D, Magill C, Hunter DA, Tong AY, Parsadanian A, Mackinnon SE, Myckatyn TM. Axotomy or compression is required for axonal sprouting following end-to-side neurorrhaphy. Exp Neurol. 2008 Jun;211(2):539–50.

[67] Viterbo F, Amr AH, Stipp EJ, Reis FJ. End-to-side neurorrhaphy: past, present, and future. Plast Reconstr Surg. 2009 Dec;124(6 Suppl):e351–8.

[68] Walker JC, Brenner MJ, Mackinnon SE, Winograd JM, Hunter DA. Effect of perineurial window size on nerve regeneration, blood-nerve barrier integrity, and functional recovery. J Neurotrauma. 2004 Feb;21(2):217–27.

[69] Tos P, Colzani G, Ciclamini D, Titolo P, Pugliese P, Artiaco S. Clinical applications of end-to-side neurorrhaphy: an update. BioMed Res Int. 2014;2014:646128.

[70] Kale SS, Glaus SW, Yee A, Nicoson MC, Hunter DA, Mackinnon SE, Johnson PJ. Reverse end-to-side nerve transfer: from animal model to clinical use. J Hand Surg. 2011 Oct;36(10):1631–9.e2.

[71] Farber SJ, Glaus SW, Moore AM, Hunter DA, Mackinnon SE, Johnson PJ. Supercharge nerve transfer to enhance motor recovery: a laboratory study. J Hand Surg. 2013 Mar;38(3):466–77.

[72] Raksakulkiat R, Leechavengvongs S, Malungpaishrope K, Uerpairojkit C, Witoonchart K, Chongthammakun S. Restoration of winged scapula in upper arm type brachial plexus injury: anatomic feasibility. J Med Assoc Thail Chotmaihet Thangphaet. 2009 Dec;92(Suppl 6):S244–50.

[73] Uerpairojkit C, Leechavengvongs S, Witoonchart K, Malungpaishorpe K, Raksakulkiat R. Nerve transfer to serratus anterior muscle using the thoracodorsal nerve for winged scapula in C5 and C6 brachial plexus root avulsions. J Hand Surg. 2009 Jan;34(1):74–8.

[74] Terzis JK, Kostas I. Suprascapular nerve reconstruction in 118 cases of adult posttraumatic brachial plexus. Plast Reconstr Surg. 2006 Feb;117(2):613–29.

[75] Bertelli JA, Ghizoni MF. Transfer of the accessory nerve to the suprascapular nerve in brachial plexus reconstruction. J Hand Surg. 2007 Sep;32(7):989–98.

[76] Bertelli JA, Ghizoni MF. Results of spinal accessory to suprascapular nerve transfer in 110 patients with complete palsy of the brachial plexus. J Neurosurg Spine. 2016 Jun;24(6):990–5.

[77] Colbert SH, Mackinnon S. Posterior approach for double nerve transfer for restoration of shoulder function in upper brachial plexus palsy. Hand (NY). 2006 Dec;1(2):71–7.

[78] Lu J, Xu J, Xu W, Xu L, Fang Y, Chen L, Gu Y. Combined nerve transfers for repair of the upper brachial plexus injuries through a posterior approach. Microsurgery. 2012 Feb;32(2):111–7.

[79] Lurje A. Concerning surgical treatment of traumatic injury of the upper division of the brachial plexus (Erb's-type). Ann Surg. 1948 Feb;127(2):317–26.

[80] Leechavengvongs S, Witoonchart K, Uerpairojkit C, Thuvasethakul P. Nerve transfer to deltoid muscle using the nerve to the long head of the triceps, part II: a report of 7 cases. J Hand Surg. 2003 Jul;28(4):633–8.

[81] Uerpairojkit C, Ketwongwiriya S, Leechavengvongs S, Malungpaishrope K, Witoonchart K, Mekrungcharas N, Chareonwat B, Ongsiriporn M. Surgical anatomy of the radial nerve branches to triceps muscle. Clin Anat (NY). 2013 Apr;26(3):386–91.

[82] Bertelli JA, Santos MA, Kechele PR, Ghizoni MF, Duarte H. Triceps motor nerve branches as a donor or receiver in nerve transfers. Neurosurgery. 2007 Nov;61(5 Suppl 2):333–9.

[83] Witoonchart K, Leechavengvongs S, Uerpairojkit C, Thuvasethakul P, Wongnopsuwan V. Nerve transfer to deltoid muscle using the nerve to the long head of the triceps, part I: an anatomic feasibility study. J Hand Surg. 2003 Jul;28(4):628–32.

[84] Jerome JTJ. Anterior deltopectoral approach for axillary nerve neurotisation. J Orthop Surg Hong Kong. 2012 Apr;20(1):66–70.

[85] Bhandari PS, Deb P. Posterior approach for both spinal accessory nerve to suprascapular nerve and triceps branch to axillary nerve for upper plexus injuries. J Hand Surg. 2013 Jan;38(1):168–72.

[86] McRae MC, Borschel GH. Transfer of triceps motor branches of the radial nerve to the axillary nerve with or without other nerve transfers provides antigravity shoulder abduction in pediatric brachial plexus injury. Hand (NY). 2012 Jun;7(2):186–90.

[87] Bertelli JA, Ghizoni MF. Nerve transfer from triceps medial head and anconeus to deltoid for axillary nerve palsy. J Hand Surg. 2014 May;39(5):940–7.

[88] Yamada S, Peterson GW, Soloniuk DS, Will AD. Coaptation of the anterior rami of C-3 and C-4 to the upper trunk of the brachial plexus for cervical nerve root avulsion. J Neurosurg. 1991 Feb;74(2):171–7.

[89] Lykissas MG, Kostas-Agnantis IP, Korompilias AV, Vekris MD, Beris AE. Use of intercostal nerves for different target neurotization in brachial plexus reconstruction. World J Orthop. 2013 Jul 18;4(3):107–11.

[90] Chu B, Hu S, Chen L, Song J. Anatomic study on intercostal nerve transfer to suprascapular nerve. Zhongguo Xiu Fu Chong Jian Wai Ke Za Zhi Zhongguo Xiufu Chongjian Waike Zazhi Chin J Reparative Reconstr Surg. 2012 Sep;26(9):1095–7.

[91] Durand S, Oberlin C, Fox M, Diverrez JP, Dauge MC. Transfer of the first intercostal nerve to supra- and infraspinatus muscles: an anatomical study and report of the first case. J Hand Surg Eur. 2009 Apr;34(2):196–200.

[92] Malungpaishrope K, Leechavengvongs S, Witoonchart K, Uerpairojkit C, Boonyalapa A, Janesaksrisakul D. Simultaneous intercostal nerve transfers to deltoid and triceps muscle through the posterior approach. J Hand Surg. 2012 Apr;37(4):677–82.

[93] Ray WZ, Murphy RKJ, Santosa K, Johnson PJ, Mackinnon SE. Medial pectoral nerve to axillary nerve neurotization following traumatic brachial plexus injuries: indications and clinical outcomes. Hand (NY). 2012 Mar;7(1):59–65.

[94] David S, Balaguer T, Baque P, Lebreton E. Transfer of pectoral nerves to suprascapular and axillary nerves: an anatomic feasibility study. J Hand Surg. 2010 Jan;35(1):92–6.

[95] Narakas AO, Hentz VR. Neurotization in brachial plexus injuries. Indication and results. Clin Orthop. 1988 Dec;(237):43–56.

[96] Chuang DC, Lee GW, Hashem F, Wei FC. Restoration of shoulder abduction by nerve transfer in avulsed brachial plexus injury: evaluation of 99 patients with various nerve transfers. Plast Reconstr Surg. 1995 Jul;96(1):122–8.

[97] Samardzic M, Antunovic V, Joksimovic M, Bacetic D. Donor nerves in the reinnervation of brachial plexus. Neurol Res. 1986 Jun;8(2):117–22.

[98] Goubier J-N, Teboul F. Transfer of the rhomboid nerve to the suprascapular nerve: An anatomical feasibility study. Chir Main. 2015 Sep;34(4):182–5.

[99] Gu Y, Xu J, Chen L, Wang H, Hu S. Long term outcome of contralateral C7 transfer: a report of 32 cases. Chin Med J (Engl). 2002 Jun;115(6):866–8.

[100] Al-Thunyan A, Al-Qattan MM, Al-Meshal O, Al-Husainan H, Al-Assaf A. Hemi-hypoglossal nerve transfer for obstetric brachial plexus palsy: report of 3 cases. J Hand Surg. 2015 Mar;40(3):448–51.

[101] Ray WZ, Pet MA, Yee A, Mackinnon SE. Double fascicular nerve transfer to the biceps and brachialis muscles after brachial plexus injury: clinical outcomes in a series of 29 cases. J Neurosurg. 2011 Jun;114(6):1520–8.

[102] Liverneaux PA, Diaz LC, Beaulieu J-Y, Durand S, Oberlin C. Preliminary results of double nerve transfer to restore elbow flexion in upper type brachial plexus palsies. Plast Reconstr Surg. 2006 Mar;117(3):915–9.

[103] Carlsen BT, Kircher MF, Spinner RJ, Bishop AT, Shin AY. Comparison of single versus double nerve transfers for elbow flexion after brachial plexus injury. Plast Reconstr Surg. 2011 Jan;127(1):269–76.

[104] Martins RS, Siqueira MG, Heise CO, Foroni L, Teixeira MJ. A prospective study comparing single and double fascicular transfer to restore elbow flexion after brachial plexus injury. Neurosurgery. 2013 May;72(5):709–14; discussion 714–715; quiz 715.

[105] Chuang DC, Yeh MC, Wei FC. Intercostal nerve transfer of the musculocutaneous nerve in avulsed brachial plexus injuries: evaluation of 66 patients. J Hand Surg. 1992 Sep;17(5):822–8.

[106] Xiao C, Lao J, Wang T, Zhao X, Liu J, Gu Y. Intercostal nerve transfer to neurotize the musculocutaneous nerve after traumatic brachial plexus avulsion: a comparison of two, three, and four nerve transfers. J Reconstr Microsurg. 2014 Jun;30(5):297–304.

[107] Moiyadi AV, Devi BI, Nair KPS. Brachial plexus injuries: outcome following neurotization with intercostal nerve. J Neurosurg. 2007 Aug;107(2):308–13.

[108] Blaauw G, Slooff ACJ. Transfer of pectoral nerves to the musculocutaneous nerve in obstetric upper brachial plexus palsy. Neurosurgery. 2003 Aug;53(2):338–41, 342.

[109] Samardzic M, Rasulic LG, Grujicic DM, Bacetic DT, Milicic BR. Nerve transfers using collateral branches of the brachial plexus as donors in patients with upper palsy—thirty years' experience. Acta Neurochir (Wien). 2011 Oct;153(10):2009–19; discussion 2019.

[110] Novak CB, Mackinnon SE, Tung THH. Patient outcome following a thoracodorsal to musculocutaneous nerve transfer for reconstruction of elbow flexion. Br J Plast Surg. 2002 Jul;55(5):416–9.

[111] Soldado F, Ghizoni MF, Bertelli J. Thoracodorsal nerve transfer for triceps reinnervation in partial brachial plexus injuries. Microsurgery. 2016 Mar;36(3):191–7.

[112] Samii A, Carvalho GA, Samii M. Brachial plexus injury: factors affecting functional outcome in spinal accessory nerve transfer for the restoration of elbow flexion. J Neurosurg. 2003 Feb;98(2):307–12.

[113] Songcharoen P, Mahaisavariya B, Chotigavanich C. Spinal accessory neurotization for restoration of elbow flexion in avulsion injuries of the brachial plexus. J Hand Surg. 1996 May;21(3):387–90.

[114] Gu YD, Ma MK. Use of the phrenic nerve for brachial plexus reconstruction. Clin Orthop. 1996 Feb;(323):119–21.

[115] Flores LP. Results of surgical techniques for re-innervation of the triceps as additional procedures for patients with upper root injuries. J Hand Surg Eur Vol. 2013 Mar;38 (3):248–56.

[116] Bertelli JA, Ghizoni MF, Tacca CP. Transfer of the teres minor motor branch for triceps reinnervation in tetraplegia. J Neurosurg. 2011 May;114(5):1457–60.

[117] Klika BJ, Spinner RJ, Bishop AT, Kircher MF, Shin AY. Posterior branch of the axillary nerve transfer to the lateral triceps branch for restoration of elbow extension: case report. J Hand Surg. 2013 Jun;38(6):1145–9.

[118] Gao K, Lao J, Zhao X, Gu Y. Outcome after transfer of intercostal nerves to the nerve of triceps long head in 25 adult patients with total brachial plexus root avulsion injury. J Neurosurg. 2013 Mar;118(3):606–10.

[119] Goubier J-N, Teboul F, Khalifa H. Reanimation of elbow extension with intercostal nerves transfers in total brachial plexus palsies. Microsurgery. 2011 Jan;31(1):7–11.

[120] Flores LP. Reanimation of elbow extension with medial pectoral nerve transfer in partial injuries to the brachial plexus. J Neurosurg. 2013 Mar;118(3):588–93.

[121] Flores LP, Socolovsky M. Phrenic nerve transfer for reconstruction of elbow extension in severe brachial plexus injuries. J Reconstr Microsurg. 2016 Sep;32(7):546–50.

[122] Zhang C-G, Gu Y-D. Contralateral C7 nerve transfer—our experiences over past 25 years. J Brachial Plex Peripher Nerve Inj. 2011 Nov 23;6(1):10.

[123] Flores LP. The radio-radial nerve transfer for elbow extension restoration in C5 to C7 nerve root injury. Microsurgery. 2012 Jan;32(1):55–9.

[124] Asfazadourian H, Tramond B, Dauge MC, Oberlin C. Morphometric study of the upper intercostal nerves: practical application for neurotizations in traumatic brachial plexus palsies. Chir Main. 1999;18(4):243–53.

[125] Luo P-B, Chen L, Zhou C-H, Hu S-N, Gu Y-D. Results of intercostal nerve transfer to the musculocutaneous nerve in brachial plexus birth palsy. J Pediatr Orthop. 2011 Dec;31(8):884–8.

[126] Pondaag W, Malessy MJA. Intercostal and pectoral nerve transfers to re-innervate the biceps muscle in obstetric brachial plexus lesions. J Hand Surg Eur Vol. 2014 Jul;39(6):647–52.

[127] Jiang Y, Wang L, Lao J, Zhao X. Comparative study of intercostal nerve transfer to lower trunk and contralateral C7 root transfer in repair of total brachial plexus injury in rats. J Plast Reconstr Aesthetic Surg JPRAS. 2016 May;69(5):623–8.

[128] Giddins GE, Kakkar N, Alltree J, Birch R. The effect of unilateral intercostal nerve transfer upon lung function. J Hand Surg Edinb Scotl. 1995 Oct;20(5):675–6.

[129] Xu W-D, Gu Y-D, Xu J-G, Tan L-J. Full-length phrenic nerve transfer by means of video-assisted thoracic surgery in treating brachial plexus avulsion injury. Plast Reconstr Surg. 2002 Jul;110(1):104–11.

[130] Zheng M-X, Qiu Y-Q, Xu W-D, Xu J-G. Long-term observation of respiratory function after unilateral phrenic nerve and multiple intercostal nerve transfer for avulsed brachial plexus injury. Neurosurgery. 2012 Apr;70(4):796–801; discussion 801.

[131] Wang S, Yiu H-W, Li P, Li Y, Wang H, Pan Y. Contralateral C7 nerve root transfer to neurotize the upper trunk via a modified prespinal route in repair of brachial plexus avulsion injury. Microsurgery. 2012 Mar;32(3):183–8.

[132] Vanaclocha V, Herrera JM, Verdu-Lopez F, Gozalbes L, Sanchez-Pardo M, Rivera M, Martinez-Gomez D, Mayorga JD. Transdiscal C6-C7 contralateral C7 nerve root transfer in the surgical repair of brachial plexus avulsion injuries. Acta Neurochir (Wien). 2015 Dec;157(12):2161–7.

[133] Mackinnon SE, Roque B, Tung TH. Median to radial nerve transfer for treatment of radial nerve palsy. Case report. J Neurosurg. 2007 Sep;107(3):666–71.

[134] Ukrit A, Leechavengvongs S, Malungpaishrope K, Uerpairojkit C, Chongthammakun S, Witoonchart K. Nerve transfer for wrist extension using nerve to flexor digitorum superficialis in cervical 5, 6, and 7 root avulsions: anatomic study and report of two cases. J Hand Surg. 2009 Nov;34(9):1659–66.

[135] Brown JM, Tung THH, Mackinnon SE. Median to radial nerve transfer to restore wrist and finger extension: technical nuances. Neurosurgery. 2010 Mar;66(3 Suppl Operative):75–83; discussion 83.

[136] Bertelli JA, Tacca CP, Winkelmann Duarte EC, Ghizoni MF, Duarte H. Transfer of the pronator quadratus motor branch for wrist extension reconstruction in brachial plexus palsy. Plast Reconstr Surg. 2012 Dec;130(6):1269–78.

[137] Bertelli JA, Kechele PR, Santos MA, Besen BAMP, Duarte H. Anatomical feasibility of transferring supinator motor branches to the posterior interosseous nerve in C7-T1 brachial plexus palsies. Laboratory investigation. J Neurosurg. 2009 Aug;111(2):326–31.

[138] Bertelli JA, Ghizoni MF. Transfer of supinator motor branches to the posterior interosseous nerve in C7-T1 brachial plexus palsy. J Neurosurg. 2010 Jul;113(1):129–32.

[139] Dong Z, Gu Y-D, Zhang C-G, Zhang L. Clinical use of supinator motor branch transfer to the posterior interosseous nerve in C7-T1 brachial plexus palsies. J Neurosurg. 2010 Jul;113(1):113–7.

[140] Palazzi S, Palazzi J-L, Caceres J-P. Neurotization with the brachialis muscle motor nerve. Microsurgery. 2006;26(4):330–3.

[141] Zheng X, Hou C, Gu Y, Shi Q, Guan S. Repair of brachial plexus lower trunk injury by transferring brachialis muscle branch of musculocutaneous nerve: anatomic feasibility and clinical trials. Chin Med J (Engl). 2008 Jan 20;121(2):99–104.

[142] Hsiao EC, Fox IK, Tung TH, Mackinnon SE. Motor nerve transfers to restore extrinsic median nerve function: case report. Hand (NY). 2009 Mar;4(1):92–7.

[143] Ray WZ, Yarbrough CK, Yee A, Mackinnon SE. Clinical outcomes following brachialis to anterior interosseous nerve transfers. J Neurosurg. 2012 Sep;117(3):604–9.

[144] García-López A, Navarro R, Martinez F, Rojas A. Nerve transfers from branches to the flexor carpi radialis and pronator teres to reconstruct the radial nerve. J Hand Surg. 2014 Jan;39(1):50–6.

[145] Vernadakis AJ, Humphreys DB, Mackinnon SE. Distal anterior interosseous nerve in the recurrent motor branch graft for reconstruction of a median nerve neuroma-in-continuity. J Reconstr Microsurg. 2004 Jan;20(1):7–11.

[146] Wang Y, Zhu S. Transfer of a branch of the anterior interosseous nerve to the motor branch of the median nerve and ulnar nerve. Chin Med J (Engl). 1997 Mar;110(3):216–9.

[147] Mackinnon SE, Yee A, Ray WZ. Nerve transfers for the restoration of hand function after spinal cord injury. J Neurosurg. 2012 Jul;117(1):176–85.

[148] Murphy RKJ, Ray WZ, Mackinnon SE. Repair of a median nerve transection injury using multiple nerve transfers, with long-term functional recovery. J Neurosurg. 2012 Nov;117(5):886–9.

[149] Bertelli JA, Mendes Lehm VL, Tacca CP, Winkelmann Duarte EC, Ghizoni MF, Duarte H. Transfer of the distal terminal motor branch of the extensor carpi radialis brevis to the nerve of the flexor pollicis longus: an anatomic study and clinical application in a tetraplegic patient. Neurosurgery. 2012 Apr;70(4):1011–6; discussion 1016.

[150] Boutros S, Nath RK, Yüksel E, Weinfeld AB, Mackinnon SE. Transfer of flexor carpi ulnaris branch of the ulnar nerve to the pronator teres nerve: histomorphometric analysis. J Reconstr Microsurg. 1999 Feb;15(2):119–22.

[151] Gaul JS. Intrinsic motor recovery—a long-term study of ulnar nerve repair. J Hand Surg. 1982 Sep;7(5):502–8.

[152] Haase SC, Chung KC. Anterior interosseous nerve transfer to the motor branch of the ulnar nerve for high ulnar nerve injuries. Ann Plast Surg. 2002 Sep;49(3):285–90.

[153] Battiston B, Lanzetta M. Reconstruction of high ulnar nerve lesions by distal double median to ulnar nerve transfer. J Hand Surg. 1999 Nov;24(6):1185–91.

[154] Barbour J, Yee A, Kahn LC, Mackinnon SE. Supercharged end-to-side anterior interosseous to ulnar motor nerve transfer for intrinsic musculature reinnervation. J Hand Surg. 2012 Oct;37(10):2150–9.

[155] Tung TH, Barbour JR, Gontre G, Daliwal G, Mackinnon SE. Transfer of the extensor digiti minimi and extensor carpi ulnaris branches of the posterior interosseous nerve to restore intrinsic hand function: case report and anatomic study. J Hand Surg. 2013 Jan;38(1):98–103.

[156] Leechavengvongs S, Ngamlamiat K, Malungpaishrope K, Uerpairotkit C, Witoonchart K, Kulkittiya S. End-to-side radial sensory to median nerve transfer to restore sensation and relieve pain in C5 and C6 nerve root avulsion. J Hand Surg. 2011 Feb;36(2):209–15.

[157] Bertelli JA, Ghizoni MF. Nerve transfer for sensory reconstruction of C8-T1 dermatomes in tetraplegia. Microsurgery. 2016 Nov;36(8):637–641.

[158] Bertelli JA, Ghizoni MF. Very distal sensory nerve transfers in high median nerve lesions. J Hand Surg. 2011 Mar;36(3):387–93.

[159] Ducic I, Dellon AL, Bogue DP. Radial sensory neurotization of the thumb and index finger for prehension after proximal median and ulnar nerve injuries. J Reconstr Microsurg. 2006 Feb;22(2):73–8.

[160] Xu B, Dong Z, Zhang C-G, Gu Y-D. Transfer of the radial branch of the superficial radial nerve to the sensory branch of the ulnar nerve for sensory restoration after C7-T1 brachial plexus injury. J Plast Reconstr Aesthetic Surg JPRAS. 2016 Mar;69(3):318–22.

[161] Campbell AA, Eckhauser FE, Belzberg A, Campbell JN. Obturator nerve transfer as an option for femoral nerve repair: case report. Neurosurgery. 2010 Jun;66(6 Suppl Operative):375; discussion 375.

[162] Tung TH, Chao A, Moore AM. Obturator nerve transfer for femoral nerve reconstruction: anatomic study and clinical application. Plast Reconstr Surg. 2012 Nov;130(5):1066–74.

[163] Spiliopoulos K, Williams Z. Femoral branch to obturator nerve transfer for restoration of thigh adduction following iatrogenic injury. J Neurosurg. 2011 Jun;114(6):1529–33.

[164] Giuffre JL, Bishop AT, Spinner RJ, Shin AY. Surgical technique of a partial tibial nerve transfer to the tibialis anterior motor branch for the treatment of peroneal nerve injury. Ann Plast Surg. 2012 Jul;69(1):48–53.

[165] Bodily KD, Spinner RJ, Bishop AT. Restoration of motor function of the deep fibular (peroneal) nerve by direct nerve transfer of branches from the tibial nerve: an anatomical study. Clin Anat (NY). 2004 Apr;17(3):201–5.

[166] Gousheh J, Arasteh E. Modified simple decompression of ulnar nerve in treatment of cubital tunnel syndrome. World J Plast Surg. 2012 Jan;1(1):11–5.

[167] Ray WZ, Mackinnon SE. Nerve problems in the lower extremity. Foot Ankle Clin. 2011 Jun;16(2):243–54.

[168] Ruggieri MR, Braverman AS, Bernal RM, Lamarre NS, Brown JM, Barbe MF. Reinnervation of urethral and anal sphincters with femoral motor nerve to pudendal nerve transfer. Neurourol Urodyn. 2011 Nov;30(8):1695–704.

[169] Fox IK. Nerve transfers in tetraplegia. Hand Clin. 2016 May;32(2):227–42.

[170] Brown JM. Nerve transfers in tetraplegia I: background and technique. Surg Neurol Int. 2011;2:121.

[171] Fridén J, Gohritz A. Tetraplegia management update. J Hand Surg. 2015 Dec;40(12): 2489–500.

[172] Bryden AM, Hoyen HA, Keith MW, Mejia M, Kilgore KL, Nemunaitis GA. Upper extremity assessment in tetraplegia: the importance of differentiating between upper and lower motor neuron paralysis. Arch Phys Med Rehabil. 2016 Jun;97(6 Suppl):S97–104.

[173] Hahn J, Cooper C, Flood S, Weymouth M, van Zyl N. Rehabilitation of supinator nerve to posterior interosseous nerve transfer in individuals with tetraplegia. arch phys med rehabil. 2016 Jun;97(6 Suppl):S160–8.

[174] Bertelli JA, Ghizoni MF. Nerve transfers for elbow and finger extension reconstruction in midcervical spinal cord injuries. J Neurosurg. 2015 Jan;122(1):121–7.

[175] Bertelli JA, Ghizoni MF. Nerve transfers for restoration of finger flexion in patients with tetraplegia. J Neurosurg Spine. 2017 Jan;26(1):55–61.

[176] Fox IK, Davidge KM, Novak CB, Hoben G, Kahn LC, Juknis N, Ruvinskaya R, Mackinnon SE. Use of peripheral nerve transfers in tetraplegia: evaluation of feasibility and morbidity. Hand (NY). 2015 Mar;10(1):60–7.

[177] Hawasli AH, Chang J, Reynolds MR, Ray WZ. Transfer of the brachialis to the anterior interosseous nerve as a treatment strategy for cervical spinal cord injury: technical note. Glob Spine J. 2015 Apr;5(2):110–7.

[178] van Zyl N, Hahn JB, Cooper CA, Weymouth MD, Flood SJ, Galea MP. Upper limb reinnervation in C6 tetraplegia using a triple nerve transfer: case report. J Hand Surg. 2014 Sep;39(9):1779–83.

Chapter 4

Surgical Treatment of Brachial Plexus Injury

Pavel Haninec and Libor Mencl

Additional information is available at the end of the chapter

http://dx.doi.org/10.5772/intechopen.68442

Abstract

In recent years, brachial plexus injury has been attracting increasing attention, partly because of an increasing incidence arising out of higher survival rates for patients after polytrauma. Brachial plexus injury is one of the hardest and most mutilating injuries. Owing to advances in microsurgical techniques, we can achieve success in restoring motor function for these patients. The purpose of this chapter is to introduce the reader with various microsurgical techniques, including nerve fascicle transfers and end-to-side neurorrhaphy (ETSN), which can be used for brachial plexus reconstruction based on personal experience with 1130 nerve reconstructions performed by the first author (PH) between 1993 and 2017. Another goal of brachial plexus surgery is the resolution of severe intractable pain which can develop in up to 20% of cases. Dorsal root entry zone (DREZ) thermocoagulation is a very effective method for treatment of severe neuropathic pain.

Keywords: brachial plexus surgery, neurotization, fascicular transfer, end-to-side neurorrhaphy, obstetrical brachial plexus palsy, root avulsion, DREZ thermocoagulation

1. Introduction

Recently, brachial plexus injury has been attracting increasing attention, partly because of an increasing incidence arising from higher survival rates in patients after polytrauma. The purpose of this chapter is to introduce the reader with various microsurgical techniques which can be used for brachial plexus reconstruction based on personal experience with over than 990 nerve reconstructions performed by the first author (PH) between 1993 and 2017.

Being one of the hardest and most mutilating injuries, the main goals of brachial plexus injury surgeries are the restoration of movement in the affected segments of the upper extremity and the resolution of pain. Owing to advances in microsurgical techniques, we can achieve success in restoring motor function in 60–80% patients [1, 2]. There is an absence of publications dealing with the epidemiology of brachial plexus injury. Midha described 54 cases of brachial plexus injury in 4538 patients after polytrauma (1.2%), while the total number of brachial plexus injuries in patients after winter sports injuries and motorcycle collisions was much higher, 4.8 and 4.2%, respectively [3]. The majority of injuries are upper brachial plexus palsies (72%), followed by complete lesions (26%). Lower plexus palsies were present only in 2% of the cases [4].

There are two main types and mechanisms of brachial plexus injury. The first is the traction injury, which results in the avulsion of cervical roots and the other is direct injury to the trunks, cords, and nerves of the brachial plexus [5]. The timing of surgery remains controversial mainly because of a lack of large randomized clinical trials of this specific surgical procedure. In cases of open injury, immediate neurosurgical revision is recommended within 72 h. Neurosurgical revision of closed brachial plexus injury in 3–6 months is acceptable, if there is no useful spontaneous reinnervation [1]. Treatment of patients with avulsion of cervical roots is especially difficult, since direct reconstruction is impossible and the results of reimplantation of avulsed roots are thoroughly controversial and have been presented in only a small number of cases [6].

2. Preoperative examination

Precise assessment of preoperative neurological status is highly recommended. Muscle strength in each our patient was graded, using the Medical Research Council scale: grade 0, no contraction; grade 1, flicker or trace contraction; grade 2, active movement with gravity eliminated; grade 3, active movement against gravity; grade 4, active movement against resistance; and grade 5, normal strength [7]. Sensory function was evaluated in the distribution area of the roots and peripheral nerves. Attention was paid to the presence of Tinel's sign or Horner's syndrome and detailed evaluation of pain. Clinical assessment was extended by radiological and electrophysiological examination. All the patients had undergone preoperative electromyography examination by means of needle concentric electrode and nerve conduction examinations. In addition to performing electromyography assessment of muscles innervated by the nerves of the brachial plexus, we focused on evaluation of muscles innervated by possible donors for the neurotization procedures [8]. The donors can be divided into two groups. The group of extraplexal donors which are not part of the brachial plexus: the accessory nerve (upper part of the trapezius) and motor branches of the C-4 spinal nerve (levator scapulae). And the group of intraplexal donors, direct branches of the brachial plexus: the thoracodorsal nerve (latissimus dorsi), the long thoracic nerve (serratus anterior), and the pectoral nerves (pectoralis major). Somatosensory evoked potentials were recorded in order to help distinguish the different levels of brachial plexus injury. Computed tomography myelography was considered to be the technique of choice for diagnostic root avulsions with a reported overall diagnostic accuracy of 85% [9]. This is an invasive, yet safe procedure, completed without any serious complications in all our patients. This examination is able to detect not only the presence of pseudocysts but also partial root avulsion in the absence of a rupture of the enveloping meninges [10]. RTG retinoscopy or ultrasonography is routinely used for evaluation of diaphragm function.

3. Operative techniques

3.1. Direct repair procedures

Surgery is performed under general anesthesia without the use of muscle-blocking agents. Once the plexus is exposed through anterior supraclavicular, infraclavicular, or combined approach, a visual assessment of the presence, location, and extent of the neuroma takes place. Further operation technique is selected based on preoperative knowledge of the presence of avulsions and selective electrical stimulation of all involved nerve portions. The nerve action potential recording is carried out with the aim of demonstrating the presence or the absence of regenerating myelinated fibers through the site of the lesion. Direct bipolar electrical stimulation and visual evaluation of the muscle contraction is performed as well. In cases in which visual evidence of contraction was in doubt, intraoperative electromyography evaluation of motor response was carried out using concentric needle electrodes. If a sufficient muscle response after intraoperative electrical stimulation proximal to the neuroma is present, then only neurolysis is performed. If there is no useful response to electrical stimulation, the neuroma tissue is resected in a stepwise manner. The preferred option is direct coaptation between available proximal and distal nerve stumps. If this is not possible, nerve grafts are harvested and used across the gap. In most cases sural nerves are harvested from one or both legs and then used as grafts. Removal of these nerves does not produce any severe deficits in the legs. After the nerves have been harvested, there can be a slight loss of sensation in the lateral part of the foot, which tends to diminish progressively with time. Another option is to use medial or less frequently lateral antebrachial cutaneous nerve or free vascularized noncritical injured nerve as a graft [11]. The overall success rate of direct repair was 79% in our previously presented study, with the average period leading to the first signs of reinnervation measured by EMG evaluation being 12 months [1].

3.2. Nerve transfers

Nerve transfer (neurotization) is a method using the transfer of a functional but less important donor nerve to a nonfunctional, more important recipient nerve in cases of cervical root avulsions or intractable proximal brachial plexus injury. The choice of donor nerves for neurotization remains a controversial topic in peripheral nerve surgery mainly because of a lack of large randomized clinical trials of this specific surgical procedure [12]. In general, there are two categories of donor nerves for neurotization. The use of extraplexal nerves such as an accessory nerve, motor branches of the cervical plexus, intercostal nerves, or the phrenic nerve, is the only option for patients with complete avulsion of cervical roots forming a brachial plexus. The use of intraplexal nerves as donors of motor fibers, such as the thoracodorsal nerve, the long thoracic nerve, and the pectoral nerves leads to more favorable results than the use of extraplexal donors [1]. The time interval between injury and surgery can influence surgical outcome [13]. The suprascapular nerve, the crucial nerve in shoulder abduction and external rotation, arises from the C-5 and C-6 roots and rarely from C-4 [14]. The incidence of C-4 nerve participation in the brachial plexus, as well as of the suprascapular nerve, differs between 18 and 62.8% among studies. In such cases, shoulder abduction may be stronger, especially at the beginning of the motion [15]. The suprascapular nerve is reconstructed from one of the roots or separately with the spinal accessory nerve as donor. We preserved the proximal branches to the trapezius muscle in all cases. The success rate of this procedure is 80–90%, and it is generally accepted that simultaneous neurotization of the suprascapular and axillary nerves offers better overall shoulder function [16]. In axillary nerve reconstruction, good results have been obtained using the: (1) thoracodorsal nerves (86%), (2) intercostal nerves (33–67%), (3) phrenic nerve (66%), (4) spinal accessory nerve (60%), (5) C-5 and/or C-6 root via nerve grafts (67%), or (6) the contralateral C-7 root (52%) as the donor of motor nerve fibers. [1, 2, 16–19]. Best results can be achieved in cases with a preserved proximal stump of the axillary nerve that can be reconstructed via end-to-end suturing using a short nerve graft [16]. When other donor nerves were not available, especially in total brachial plexus palsies, recently we have been using the long thoracic nerve (from C-4 to C-7). The disadvantage of this method rests with the risk of scapular instability [16]. The method of choice in axillary nerve reconstruction is, in our opinion, neurotization by using the thoracodorsal nerve, which has an 86% success rate [1]. The thoracodorsal nerve arises from C5–7 and is often available in injuries of the upper plexus [14]. So far, numerous methods of musculocutaneous nerve neurotization have been defined. Successful reinnervation with this technique has been obtained using: (1) the medial pectoral nerve (84–91%), (2) the intercostal nerves (44–70%), (3) phrenic nerve (66%), and (4) spinal accessory nerve (50–88%) as donor nerves [1, 20]. The intercostal nerves may be destructed in patients with rib fractures. In such cases, the intercostal nerves cannot be used for neurotization. We do not use the phrenic nerve as a donor because its use may result in poor outcomes and increased risk of diaphragmatic impairment [1]. Methods of successful reinnervation of radial nerve have been described using: (1) branch of the median nerve 91.7%, (2) from the dorsal cord (93–100%) [1, 11]. The least successful results were achieved after the reconstructions of the ulnar and median nerves. A long distance to the motor point of the reinnervated muscle and a long reinnervation time are the reasons for poor results [1, 21].

3.3. Fascicular transfer

Recently, new promising ways of neurotization using only a part of the donor nerve have been published. Transfer of the triceps motor branches of the radial nerve to the axillary nerve was performed to restore deltoid muscle function and appears to be safe and effective (**Figure 1**). The functional loss relative to the triceps, with a single nerve transferred, is negligible because of compensation by the remaining heads [22]. Oberlin introduced a neurotization technique in which an ulnar nerve fascicle is transferred to the branch of the musculocutaneous nerve for biceps muscle, thus reanimating flexion of elbow in patients with upper brachial plexus injuries without significant motor or sensory deficits of the donor nerve [23]. This technique can only be performed in patients in whom the lower brachial plexus is intact, as the ulnar nerve is formed from the C-8 and T-1 roots [14]. The Oberlin technique is now routinely used to reinnervate the biceps muscle and restore elbow flexion in cases of upper brachial plexus palsy (**Figure 2**). In 2001, Songcharoen described an alternative technique using a median nerve fascicle to repair the musculocutaneous nerve and restore biceps muscle function [18]. Based on these findings, we performed this neurotization technique on a group of 17 patients with median age 34 years (SD = 12.7, range 17–58), the median time between trauma and

http://dx.doi.org/10.5772/intechopen.68442

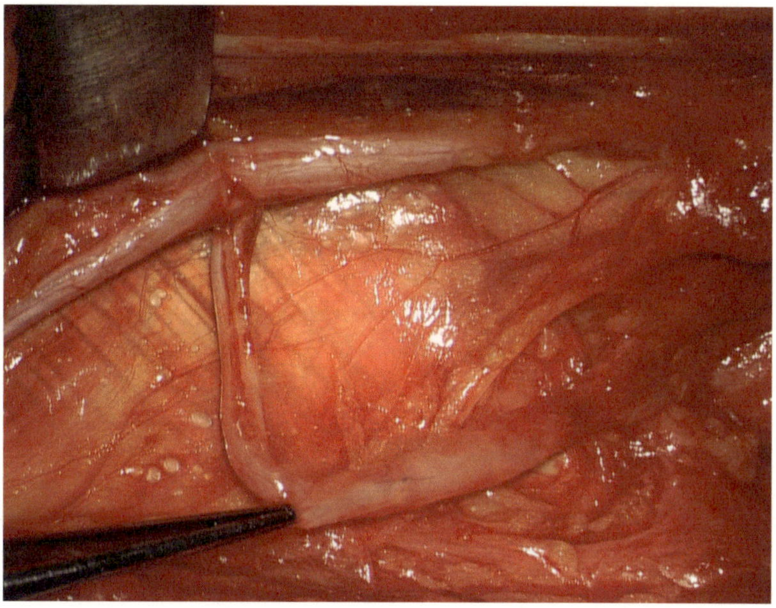

Figure 1. Transfer of the triceps motor branches of the radial nerve to the axillary nerve.

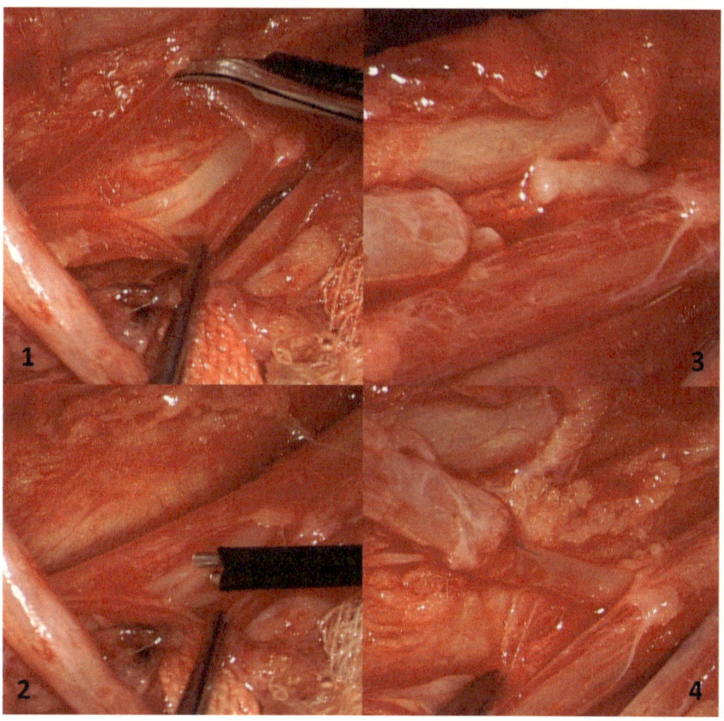

Figure 2. Fascicular transfer to the musculocutaneous nerve. (1) Epineurotomy of ulnar nerve; (2) detection of fascicle with direct bipolar electrical stimulation; (3) preparation; (4) direct coaptation.

surgery was 6 months (SD = 2.7, range 2–13). This method produces outstanding results, with success rates between 90 and 100% when the donor is the ulnar nerve and 64–80% when the donor is the median nerve. The major benefit of this technique is the option to create a distal suture close to the target muscle [18, 24, 25]. However, we do not use this particular method in musculocutaneous nerve reconstruction because we have had very positive experiences using the pectoral nerves for neurotization of this nerve [20].

In 2007, Haninec published the effectiveness of motor fascicle transfer in axillary nerve reconstruction using one or two fascicles from the ulnar or median nerve [12]. Between 2007 and 2017, the first author performed this operation in a subset of 24 patients. The median age of these patients was 29 years (SD = 13.4, range 18–61). The median time between trauma and reconstructive surgery was 5 months (SD = 2.4, range 2–13). The axillary nerve was divided 3 cm above the foramen of Velpeau and dorsal to the subscapular vessels. The fascicle of the ulnar nerve for flexor carpi ulnaris muscle was carefully selected using electrical stimulation of the proximal part of the arm. The localization of this fascicle was verified after epineurotomy through observation of the reaction of the corresponding muscle to direct bipolar electrical stimulation. In the case of the absence of muscle activity in the flexor carpi ulnaris muscle, the fascicle of the median nerve for the flexor carpi radialis muscle can be used same way. The distal portion of the axillary nerve was transferred ventrally from the subscapular vessels and medially from the radial nerve and sutured to prepared fascicle of the ulnar or median nerve (**Figure 3**). The success rate for the whole group of patients was 78.6% [12]. None of the patients lost motor function in the hand. Only two of the patients evaluated in the present study had

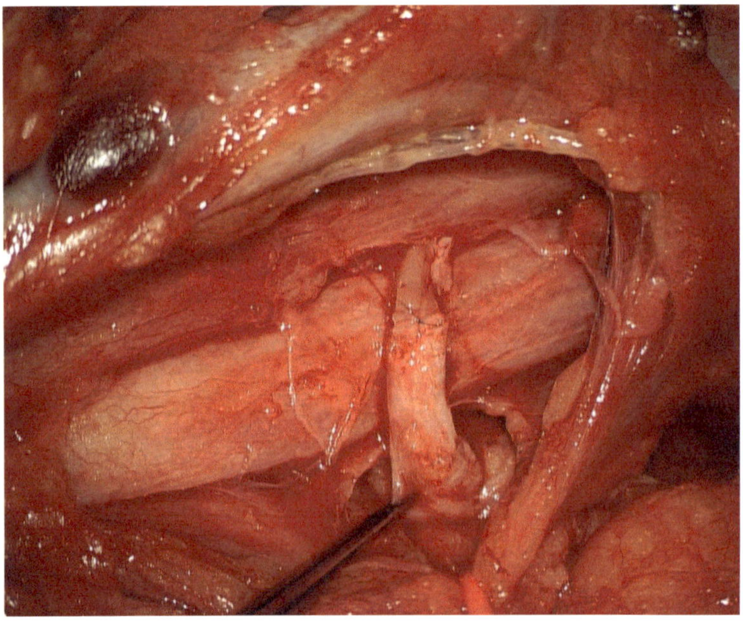

Figure 3. Ulnar fascicle transfer to the axillary nerve. The axillary nerve was transected 3 cm cranial to foramen of Velpeau and dorsal to the subscapular vessels.

a partial short-term sensory loss following the surgery. This is in concordance with the previously reported data [18, 24, 25]. The advantages of the fascicular transfer techniques include reduced regeneration distance, faster reinnervation because the nerve repair can be performed much closer to the neuromuscular junction, and also lower level of invasiveness and the fact that the nerve transfer is performed on nontraumatized tissue distal to the site of injury [12].

3.4. End-to-side neurorrhaphy

In cases when commonly used donor nerves are not available or when there is a high risk of transection of the donor nerve fascicle, end-to-side neurorrhaphy (ETSN) could be used for nerve repair in brachial plexus reconstructive surgery. This situation is expected in patients with combined supra- and infraclavicular brachial plexus injuries resulting in impaired hand functions. Advantage of ETSN over end-to-end neurotization is that with ETSN there is no need to sacrifice the surrounding nerves or their fascicles [20]. End-to-side neurorrhaphy was rediscovered in the early 1990s [26]. This technique is based on the regeneration of an injured recipient nerve through collateral sprouting of axons, one of the most important manifestations of neuroplasticity of an intact donor nerve and has been the subject of many theoretical and experimental studies [27–29]. Haninec et al. evaluated results of ETSN on a group of patients with an upper brachial injury for axillary or musculocutaneous nerve reconstruction [20]. The patients cohort consisted of 23 patients, the median age of the patients was 25 years (SD = 14.0, range 12–63), and the median time between trauma and surgery was 5 months (SD = 2.0, range 2–9). Twelve patients had cervical roots C5–6 avulsed, 11 patients had cervical roots C5–7 avulsed. The ETSN was performed through a perineurial suture of the axillary or musculocutaneous nerve onto the side of the ulnar, median, or radial nerve after the creation of a perineurial window. The success of the ETSN was achieved in 10 out of the 23 patients, corresponding to an overall rate of success of 43.5%. The choice of donor nerve (ulnar or median) did not affect the success rate. The precise site on the donor nerve where ETSN is to be performed is chosen by direct bipolar electrical stimulation and registration in the corresponding muscle. It can be found on the surface of the donor nerve from which the maximum amplitude of motor response is evoked [20]. A perineurial suture is performed after the creation of a perineurial window. This method is our standard technique because studies with experimental models have indicated that the success rate of such sutures is high [30]. In experimental models, there were no signs of denervation of the donor nerve after creation of a perineurial window [30, 31]. Similarly, none of the patients in our series had any sensory or motor loss in the innervation zone of the donor nerve. The existence of collateral sprouting of axons is well documented. Some authors' experimental studies have demonstrated that the terminolateral nerve repair can be used without a surgical incision into the donor nerve's perineurium or epineurium [32]. These methods do not seem applicable in brachial plexus surgery because only a few collateral branches grow from the mixed donor nerve with a rather thick perineurium into the mixed recipient nerve and these collateral branches are very thin. In our previous study, we observed that a perineurial suture is the only possible way to achieve effective reinnervation in nerves with large diameters such as those found in the brachial plexus [28]. Successful ETSN takes place via collateral sprouting of intact axons and also through direct growth of some injured axons into the recipient nerve. Without the contribution of the injured axons, the outcome of ESTN would be very poor, since only about 1.4% of motor neurons with intact axons can send out collateral branches. In addition, only a limited number of motor neurons have the capacity to send out collateral sprouts from intact axons [31]. Several experimental studies have suggested that sensory reinnervation is superior to motor reinnervation after ETSN. The explanation for better sensory reinnervation after ETSN may lie in the higher total number of sensory fibers than the number of motor fibers contributing to the reinnervation [32]. However, our previous results showed that motor and sensory neurons have a similar ability to send out collateral sprouts from their axons in rats with a perineurial window [28]. On the basis of our results, we recommend the use of ETSN only in cases when commonly used donor nerves are not available. Advantage of ETSN over end-to-end neurotization is that with ETSN there is no need to sacrifice the surrounding nerves or their fascicles. One possible method of getting better results in the future may be the use of neurotrophics to strengthen the reinnervation process [31].

4. Obstetrical brachial plexus injury

Obstetrical brachial plexus injury (OBPI) is caused by excessive lateral traction to the infants' head during delivery, although cases of OBPI injury after nontraumatic caesarean sections have also been described, and so perinatal brachial plexus injury would be an alternative nomenclature. Obstetrical brachial plexus palsy displays a stable incidence of 0.15–3 per 1000 live births [33]. Shoulder dystocia, macrosomia, and instrument delivery, forceps or vacuum extraction, present the greatest risk for brachial plexus injury. Caesarean section and having a twin or multiple birth mates seem to offer some protection against injury [33, 34]. Most children show good spontaneous recovery, but a recent literature review showed that a residual deficit remains in 20–30% of children [35]. The final nerve injury may vary from neurapraxia, axonotmesis, neurotmesis to cervical root avulsion from the spinal cord. Axonotmetic injuries with intact basal lamina tubes allow for axons to grow from the lesion site down into the basal lamina tube to the target muscle. Complete recovery will usually occur within the weeks or months. A more severe traction lesion results in a neurotmetic injury with a rupture of the basal lamina tubes; outgrowth of axons is blocked and does not end in appropriate endoneural tube. The consequence is formation of neuroma in continuity, a mass of outgrowing axons and scar tissue. However, even in the most severe cases, some axons will pass through and reach some tubes distal to the injury site [36]. Axons that successfully pass the neuroma in continuity are likely to end up in a different basal lamina tube from the original. This process is called misrouting and may cause the phenomenon of cocontraction [37]. Root avulsion, the most severe type of lesion, is a total disconnection between peripheral nervous system and the spinal cord. When roots are avulsed, the outgrowth of axons, neuroma formation or misrouting cannot take place. Most often, brachial plexus injuries are associated with the upper trunk, when the limb is typically held in internal rotation and pronation, with the elbow extended (roots C5–C6 ± C7, 73–86%) or complete plexus injury (roots C5–Th1, 15–20%). Isolated lower brachial plexus injury also known as Klumpke's paralysis, involving the roots of C8–Th1, does occur, but is very rare, with a frequency of only 0.16%. It is suggested that these injuries originate from complete plexus palsy with recovery of the upper part of brachial plexus [36, 38, 39]. Although we perform electromyography and imaging studies, the final decision of operation relies heavily on the clinical examination. Muscle strength testing system (MRC) although reliable for examination of motor power in adults is not suited for use with non-cooperative infants. All patients involved in our study were evaluated using the Active Movement Scale (AMS), which grades 15 upper extremity movements from 0 to 7, and greatly increases the ability to detect partial movements [40]. The results of neurophysiological investigations in older patients are mostly accurate, indicating the severity, location and extent of the lesions. In contrast, the findings of EMG and nerve-conduction studies in obstetrical brachial plexus palsy mostly suggest a falsely optimistic prognosis [41].

Between 2000 and 2016, 185 patients with obstetrical brachial plexus injury were examined at our department, and 47 underwent nerve surgery. The patient cohort consisted of 27 males and 20 females, with the right side was involved in 28 cases (59.6%). The delivery was aided with forceps or a vacuum extractor in seven cases. One case after Caesarean section was present. The patients were divided into two groups for better analysis. Group 1 involved 23 patients with upper brachial plexus birth injuries. The limb was typically held in internal rotation and pronation, with the elbow extended. An additional lack of elbow, wrist and finger extension was present in the case of root C7 involved. The median birth weight was 3840 g (range 2240–4600 g), AMS score 65 (range 15–94), clavicle fractures were present in 13% and the median Apgar score was 7–8–9. Group 2 consisted of 24 patients with complete paralysis; the upper limb was flailing, without any tonus and without hand function. The median birth weight was 4080 g (range 2550–5200 g), AMS score 3.5 (range 0– 30), clavicle fractures were present in 29% and the median Apgar score was 5–8–9. Computed tomography myelography was performed for diagnostic root avulsions. Seventy-one roots were found to be avulsed (C5, 8.5%; C6, 21.1%; C7, 23.9%; C8, 25.4%; T1, 21.1%). In most cases, the lower roots were affected; avulsion of all five roots was not seen.

Although the indication and timing of surgery remains controversial among authors [38, 42, 43], we recommend surgery in cases of total palsy with impaired hand function within the first 3 months. In cases of upper brachial plexus injuries, surgery is acceptable within the first 4 months if there is no useful elbow flexion. Operation strategy is based on preoperative knowledge of the presence of avulsions and selective intraoperative electrical stimulation of all involved nerve portions. A supraclavicular incision is sufficient for access to all roots and trunks of the brachial plexus in most cases. Infraclavicular or combined approaches are used in the case of distal nerve transfer. Neuroma resection and nerve grafting was the most frequently used technique. Nerve transfer or fascicular transfer was performed in standard manner. When the number of proximal stumps was limited in the case of root avulsion, a reconstruction of the entire brachial plexus from a single root stump was performed (**Figure 4**) [44].

The surgical objective in cases of upper OBPI is functional restoration of shoulder abduction with external rotation and biceps function. Our series reached the overall success rate 80% for shoulder abduction, 50% for external rotation and 81.8% for elbow flexion with median follow-ups of 36 months. The results of global shoulder recovery and elbow flexion [36, 38, 44, 45]. In contrast, the results with regard to external rotation are poor [46]. No difference in

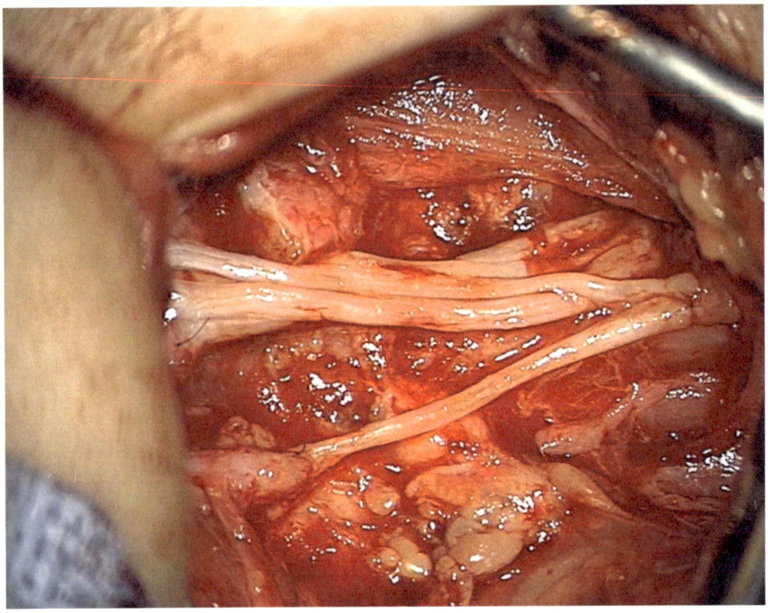

Figure 4. Neurotization following neuroma resection performed from C5 and C6 root stumps to suprascapular nerve and superior, medius, and inferior trunk with sural nerves used as grafts. C7-Th1 roots avulsed.

final external rotation was found between patients who underwent nerve grafting from the proximal stump of C5 and patients who underwent spinal accessory nerve transfer [47]. In cases of conducting neuroma, Lin et al. documented the superiority of long-term functional results following excision and grafting compared to neurolysis [48]. Reanimation of the shoulder and elbow can also be performed with nerve transfers. The use of pectoral nerve to musculocutaneous nerve neurotization and fascicular transfer using median or ulnar nerve has been described with excellent results [36, 44]. The transfer of triceps motor branches of the radial nerve to the axillary nerve appears to be effective and safe. The functional loss relative to the triceps muscle, with a one branch transferred, is meaningless because of compensation by the remaining heads of triceps muscle [49]. The surgical objective in cases with complete OBPI is recovery of hand function and establishing the ability to use the affected hand in bimanual activities [36, 42]. Bimanual execution of daily activities requires strong finger flexion in combination with good elbow flexion [50]. Reanimation of the hand is crucial; otherwise, the maximal function attained for the affected upper extremity is as a hook [36]. Recovery of the hand is dependent on the time of surgery—the younger the patient's age, the better the outcome [51]. We achieved useful hand reanimation in 87% of cases. Shoulder abduction was successful in 87%, external rotation in 25%, elbow flexion in 75%, and supination in 25% of cases. Elbow extension was successful in 87%, wrist extension in 50% and finger, thumb extension in 37% of cases with median follow-ups of 46 months (**Figure 5**) [44]. Hand function outcomes in published reports are difficult to interpret, given the variety of evaluation scales used for assessment, thus the overall success is ranging from 69 to 93% [38]. Emphasis must be placed on reconstructing the median nerve (mainly controlled by the C8 root) for innervating extrinsic

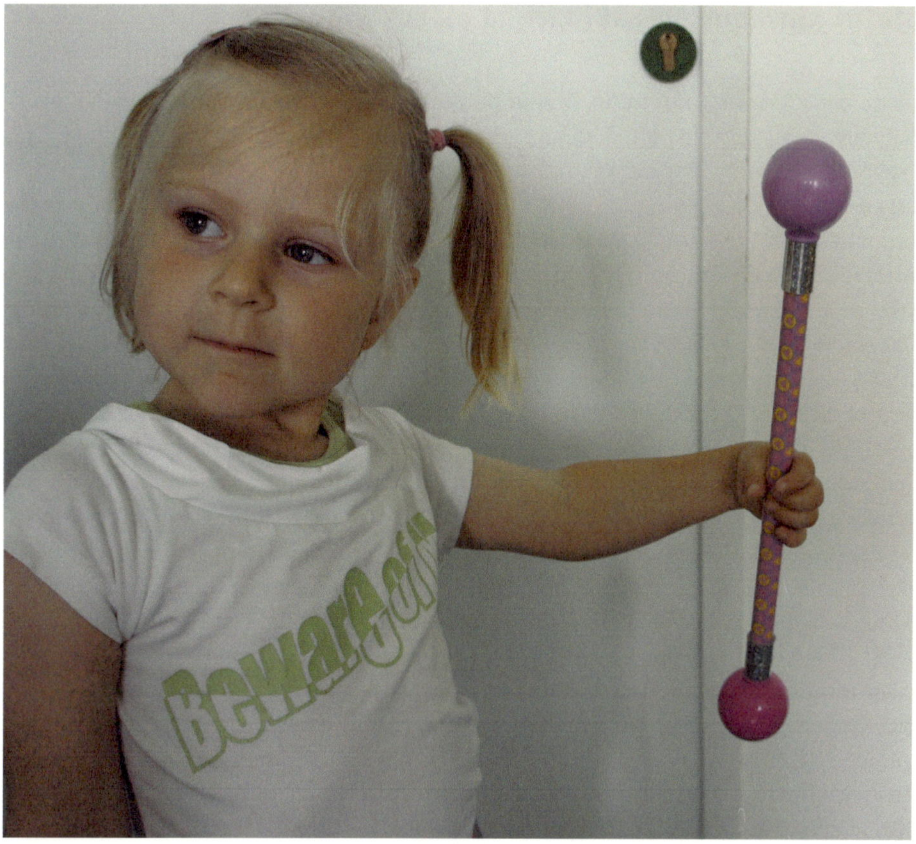

Figure 5. Result 36 months after brachial plexus reconstruction presented in Figure 4 with successful reanimation of hand.

flexor muscles of the fingers and opposition of thumb rather than the ulnar nerve innervating intrinsic muscles (mainly controlled by the Th1 root). In addition, because of the long reinnervation trajectory, severe muscle atrophy can occur [36]. The use of the end-to-side technique has been abandoned by most surgeons due to insufficient results. We described two cases with surprisingly good results, which can be attributed exclusively to the ETSN technique. In both patients we decided for a perineural suture after making a perineural window [20, 44]. This result, although with only two cases presented, may be explained by the superior nerve regeneration capacity in infants compared with adults. Laboratory reports showed better results after ETSN in young rats than in older animals [52]. In cases of late obstetrical brachial plexus injury, nerve reconstruction will be ineffective, and thus secondary surgery is called for to improve the shoulder, elbow, forearm, and hand function and to stop muscle contractures and bone shortening. The procedure includes tendon transfers, free-muscle transplantation, surgical release of muscle contractures and osteotomies. Surgery should be carried out after the age of 4 years when the child is able to cooperate with rehabilitation and on condition that severe contractures are absent [44]. Our results demonstrate that improved hand function can be obtained in infants with obstetrical brachial plexus injury with early surgical reconstruction.

5. Painful consequences of brachial plexus injury

Pain is an early symptom in up to 70% of patients with brachial plexus injury. In up to 20% of cases, severe intractable pain develops. Pain does not appear at the time of injury, but typically several days after. Typical is persistent pain with sporadic acral irritations described by patients as cutting or burning [53]. Pathophysiology of the formation is not fully explained, but it initiates after the loss of sensory impulses from the periphery which leads to the creation of pathologic pain generator in the dorsal horn of the spinal cord, in Rexed's lamina I [54]. In 90%, the pain corresponds to the avulsion of one of the lower roots. Pharmacotherapy may give satisfactory relief. However, if conservative treatment is inadequate and pain progresses, it indicates that a central component is present. The only causal therapy in these situations is dorsal root entry zone (DREZ) thermocoagulation [55]. This technique was first described by Nashold [56]. Procedure is generally performed in semisitting position with the head fixed in three-point fixation. An incision is made vertically from external occipital protuberance to the vertebra prominens. Multilevel hemilaminectomies are performed to expose the spinal cord at the level of the nerve root avulsion. The lesions are made in dorsal root entry zone in a 2-mm depth using a radiofrequency electrode. The tip of the electrode is strictly perpendicular to the surface of the spinal cord surface. Lesioning time is 15 s and lesioning temperature 75°C [55]. The risk of this procedure involves potentially serious neurological complications. The close location of thermocoagulation site from the corticospinal tract laterally and lemniscal tract dorsomedially creates a risk of motor or sensitivity failure from the point of damage. Anatomical and functional localization of DREZ is therefore essential. Problems tend to occur in cases with the dural scarring and the presence of pseudomeningocele [57].They can cause changes in spinal anatomical arrangement of the surface. A very useful technique in guiding the placement of the thermocoagulation lesions is a localization of the dorsal root entry zone by evoked potentials. A bipolar registration of evoked response and using very low intensity of stimulus (0.1–0.2mA), even in spinal cord extensively encased in scar, allows obtaining a well-differentiated response without movement artifacts. The strip electrode with two active members (5 mm distance, 1 mm in diameter) is slipped under the dura at the rostral end of exposed spinal cord. Responses are amplified at a gain setting of 100 with the high frequency filter set at 5 kHz and the low frequency filter set at 20 Hz. Responses are not averaged. The stimulating electrode is a bipolar stimulating electrode with constant distance between tips of 1 mm. A 200-μs square-wave impulse and a stimulation rate of three stimuli per second are used. Initially, the stimulating electrode is placed over the dorsal column and the intensity of the stimulus is gradually increased until an evoked potential with amplitude of approximately 30–50 uV is elicited. The stimulus intensity is not changed during the rest of the procedure and varies from 0.1 to 0.2 mA. Dorsolateral surface of spinal cord is gradually stimulated at a constant distance from the registration electrode, in approximately 2 mm steps. Sites where the stimulation electrode failed to evoke a response are considered as a dorsal root entry zone and thermocoagulation lesion is made at that site. The stimulating technique is repeated along the axis of the spinal cord at 1 cm intervals [58]. Surgical procedures were performed by the first author between 1993 and 2016 on 61 patients. The patient cohort consisted of 56 men and 5 women with a median age of 38 years (SD = 11.4, range 20–70). Median number of performed thermo-lesions were 29 (SD = 9.2, range 13–50). A decrease in preoperative pain intensity of more than 75% was considered a definite success. Based on our long-term follow-up (median follow-up 61 months, range 15–180), this goal was achieved in 68.4% of patients. Another 22.8% reported decrease of pain intensity between 75 and 50% and reported some pain persistence, usually in the form of dull pain or paresthesias of the affected upper limb. Overall satisfaction with the surgery was achieved in 91.2% of patients. An operative painkilling surgery is generally required in 10–15% of patients with spinal cord root avulsion [59, 60]. The incidence of complications is different for different authors, ranging from 0 to 60% [57]. Haninec et al. found in a group of 48 patients that frequency of complications was 15.4%, sensory deficits occurred in five cases (1× hemihypoesthesia of the trunk and lower limb, 4× hypoesthesia of ipsilateral lower extremity), motor leg weakness occurred in two patients and one case had a combined disability [55]. The best results have been achieved in sporadic irritations while persistent dull pain had a worse prognosis and higher tendency to recur [53, 61]. Paroxysmal pain is successfully eliminated in 91.6% while severe dull pain was treatable in 70.8% cases [55]. It has been stated that there is no correlation between the number of roots avulsed or the extent of the DREZ T procedure performed and the degree of pain reduction [60]. DREZ thermocoagulation is a very effective method for treatment of severe neuropathic pain that can develop in some patients with supraganglion brachial plexus injury. Needless to say, such delicate procedures require masterful execution.

6. Conclusions

Brachial plexus injury is one the hardest and most mutilating injuries. Owing to advances in microsurgical techniques, we can achieve success in restoring motor function for these patients. Direct repair and neurotization has become the mainstay for the brachial plexus surgery. The presence and following choice of donor nerves for neurotization is most important. Using only a part of the donor nerve for neurotization is a new promising technique. The advantages of the fascicular transfer include reduced regeneration distance, faster reinnervation, and lower level of invasiveness. The timing of surgery remains controversial mainly because of a lack of large randomized clinical trials of this specific surgical procedure. The use of end-to-side neurorrhaphy is recommended only in cases when commonly used donor nerves are not available. In cases of obstetrical brachial plexus injury, shoulder dystocia, macrosomia, and instrument delivery present the greatest risk. Our results demonstrate that improved hand function can be obtained in infants with early surgical reconstruction. One possible method of getting better results in the future may be the use of neurotrophics to strengthen the reinnervation process. Dealing with painful consequences of brachial plexus injury, DREZ thermocoagulation is a very effective method for the treatment of severe neuropathic pain.

Acknowledgements

This study was supported by Charles University in Prague, Progress Q35.

Author details

Pavel Haninec* and Libor Mencl

*Address all correspondence to: haninec@fnkv.cz

Department of Neurosurgery, 3rd Faculty of Medicine, Charles University, Prague, Czech Republic

References

[1] Haninec P, et al. Direct repair (nerve grafting), neurotization, and end-to-side neurorrhaphy in the treatment of brachial plexus injury. Journal of Neurosurgery. 2007;**106**(3):391–399

[2] Kim DH, et al. Outcomes of surgery in 1019 brachial plexus lesions treated at Louisiana State University Health Sciences Center. Journal of Neurosurgery. 2003; **98**(5):1005–1016

[3] Midha R. Epidemiology of brachial plexus injuries in a multitrauma population. Neurosurgery. 1997;**40**(6):1182–1189

[4] Kaiser R, Mencl L, Haninec P. Injuries associated with serious brachial plexus involvement in polytrauma among patients requiring surgical repair. Injury. 2014; **45**(1):223–226

[5] Kline DG, Hudson AR. Nerve Injuries: Operative Results For Major Nerve Injuries, Entrapments, and Tumors. Philadelphia: Saunders; 1995

[6] Carlstedt T, et al. Spinal nerve root repair and reimplantation of avulsed ventral roots into the spinal cord after brachial plexus injury. Journal of Neurosurgery: Spine. 2000;**93**(2):237–247

[7] Seddon H. Peripheral nerve injuries. Medical Research Council Special Report Series No. 282. London: Her Majesty's Stationary Office; 1954. pp. 1–7

[8] Kline DG, Hackett ER, May PR. Evaluation of nerve injuries by evoked potentials and electromyography. Journal of Neurosurgery. 1969;**31**(2):128–136

[9] Carvalho GA, et al. Diagnosis of root avulsions in traumatic brachial plexus injuries: Value of computerized tomography myelography and magnetic resonance imaging. Journal of Neurosurgery. 1997;**86**(1):69–76

[10] Steens SC, et al. Obstetric brachial plexus lesions: CT myelography. Radiology. 2011;**259**(2):508–515

[11] Davidge KM, et al. Median to radial nerve transfers for restoration of wrist, finger, and thumb extension. The Journal of Hand Surgery. 2013;**38**(9):1812–1827

[12] Haninec P, Kaiser R. Axillary nerve repair by fascicle transfer from the ulnar or median nerve in upper brachial plexus palsy: Clinical article. Journal of Neurosurgery. 2012;**117**(3):610–614

[13] Samii A, Carvalho GA, Samii M. Brachial plexus injury: Factors affecting functional outcome in spinal accessory nerve transfer for the restoration of elbow flexion. Journal of Neurosurgery. 2003;**98**(2):307–312

[14] Bonnel F. Microscopic anatomy of the adult human brachial plexus: An anatomical and histological basis for microsurgery. Microsurgery. 1984;**5**(3):107–118

[15] Shin C, et al. Spinal root origins and innervations of the suprascapular nerve. Surgical and Radiologic Anatomy. 2010;**32**(3):235–238

[16] Vekris MD, et al. Shoulder reanimation in posttraumatic brachial plexus paralysis. Injury. 2010;**41**(3):312–318

[17] Samardzic M, et al. Results of nerve transfers to the musculocutaneous and axillary nerves. Neurosurgery. 2000;**46**(1):93–103

[18] Songcharoen P. Management of brachial plexus injury in adults. Scandinavian Journal of Surgery. 2008;**97**(4):317–323

[19] Terzis JK, Kokkalis ZT. Selective contralateral C7 transfer in posttraumatic brachial plexus injuries: A report of 56 cases. Plastic and Reconstructive Surgery. 2009;**123**(3):927–938

[20] Haninec P, Mencl L, Kaiser R. End-to-side neurorrhaphy in brachial plexus reconstruction: Clinical article. Journal of Neurosurgery. 2013;**119**(3):689–694

[21] Spinner RJ, Kline DG. Surgery for peripheral nerve and brachial plexus injuries or other nerve lesions. Muscle & Nerve, 2000;**23**(5):680–695

[22] Leechavengvongs S, et al. Nerve transfer to deltoid muscle using the nerve to the long head of the triceps, part II: A report of 7 cases. The Journal of Hand Surgery. 2003;**28**(4):633–638

[23] Oberlin C, et al. Nerve transfer to biceps muscle using a part of ulnar nerve for C5–C6 avulsion of the brachial plexus: Anatomical study and report of four cases. The Journal of Hand Surgery. 1994;**19**(2):232–237

[24] Nath RK, Lyons AB, Bietz G. Physiological and clinical advantages of median nerve fascicle transfer to the musculocutaneous nerve following brachial plexus root avulsion injury. Journal of Neurosurgery. 2006;**105**(6):830–834

[25] Tung TH, Novak CB, Mackinnon SE. Nerve transfers to the biceps and brachialis branches to improve elbow flexion strength after brachial plexus injuries. Journal of Neurosurgery. 2003;**98**(2):313–318

[26] Viterbo F, et al. End-to-side neurorrhaphy with removal of the epineurial sheath: An experimental study in rats. Plastic and Reconstructive Surgery. 1994;**94**(7):1038–1047

[27] Bajrović F, et al. Interneuronal signalling is involved in induction of collateral sprouting of nociceptive axons. Neuroscience. 2002;**111**(3):587–596

[28] Haninec P, Kaiser R, Dubový P. A comparison of collateral sprouting of sensory and motor axons after end-to-side neurorrhaphy with and without the perineurial window. Plastic and Reconstructive Surgery. 2012;**130**(3):609–614

[29] Haninec P, et al. Enhancement of musculocutaneous nerve reinnervation after vascular endothelial growth factor (VEGF) gene therapy. BMC Neuroscience. 2012; **13**(1):57

[30] Geuna S, Papalia I, Tos P. End-to-side (terminolateral) nerve regeneration: A challenge for neuroscientists coming from an intriguing nerve repair concept. Brain Research Reviews. 2006;**52**(2):381–388

[31] Dubový P, et al. Ciliary neurotrophic factor promotes motor reinnervation of the musculocutaneous nerve in an experimental model of end-to-side neurorrhaphy. BMC Neuroscience. 2011;**12**(1):58

[32] Pannucci C, et al. End-to-side nerve repair: Review of the literature. Restorative Neurology and Neuroscience. 2007;**25**(1):45–63

[33] Foad SL, Mehlman CT, Ying J. The epidemiology of neonatal brachial plexus palsy in the United States. Journal of Bone and Joint Surgery America. 2008;**90**(6):1258–1264

[34] Gilbert WM, Nesbitt TS, Danielsen B. Associated factors in 1611 cases of brachial plexus injury. Obstetrics & Gynecology. 1999;**93**(4):536–540

[35] Pondaag W, et al. Natural history of obstetric brachial plexus palsy: A systematic review. Developmental Medicine & Child Neurology. 2004;**46**(2):138–144

[36] Malessy MJ, Pondaag W. Nerve surgery for neonatal brachial plexus palsy. Journal of Pediatric Rehabilitation Medicine. 2011;**4**(2):141–148

[37] Anguelova GV, et al. A cross-sectional study of hand sensation in adults with conservatively treated obstetric brachial plexus lesion. Developmental Medicine & Child Neurology. 2013;**55**(3):257–263

[38] Gilbert A, Pivato G, Kheiralla T. Long-term results of primary repair of brachial plexus lesions in children. Microsurgery. 2006;**26**(4):334–342

[39] Al-Qattan M, Clarke H, Curtis C. Klumpke's birth palsy: Does it really exist? The Journal of Hand Surgery: British & European Volume. 1995;**20**(1):19–23

[40] Clarke HM, Curtis CG. An approach to obstetrical brachial plexus injuries. Hand Clinics. 1995;**11**(4):563–580

[41] VAN D, et al. Needle electromyography at 1 month predicts paralysis of elbow flexion at 3 months in obstetric brachial plexus lesions. Developmental Medicine & Child Neurology. 2012;**54**(8):753–758

[42] Vekris MD, et al. Management of obstetrical brachial plexus palsy with early plexus microreconstruction and late muscle transfers. Microsurgery. 2008;**28**(4):252–261

[43] El-gammal TA, et al. Total obstetric brachial plexus palsy: results and strategy of microsurgical reconstruction. Microsurgery. 2010;**30**(3):169–178

[44] Mencl L, Waldauf P, Haninec P. Results of nerve reconstructions in treatment of obstetrical brachial plexus injuries. Acta Neurochirurgica. 2015;**157**(4):673–680

[45] Birch R, et al. Repair of obstetric brachial plexus palsy. Bone & Joint Journal. 2005;**87**(8): 1089–1095

[46] Pondaag W, et al. External rotation as a result of suprascapular nerve neurotization in obstetric brachial plexus lesions. Neurosurgery. 2005;**57**(3):530–537

[47] Tse R, et al. Suprascapular nerve reconstruction in obstetrical brachial plexus palsy: Spinal accessory nerve transfer versus C5 root grafting. Plastic and Reconstructive Surgery. 2011;**127**(6):2391–2396

[48] Lin JC, et al. Final results of grafting versus neurolysis in obstetrical brachial plexus palsy. Plastic and Reconstructive Surgery. 2009;**123**(3):939–948

[49] McRae MC, Borschel GH. Transfer of triceps motor branches of the radial nerve to the axillary nerve with or without other nerve transfers provides antigravity shoulder abduction in pediatric brachial plexus injury. Hand. 2012;**7**(2):186–190

[50] Krumlinde-Sundholm L, et al. The assisting hand assessment: Current evidence of validity, reliability, and responsiveness to change. Developmental Medicine & Child Neurology. 2007;**49**(4):259–264

[51] Terzis JK, Kokkalis ZT. Outcomes of hand reconstruction in obstetric brachial plexus palsy. Plastic and Reconstructive Surgery. 2008;**122**(2):516–526

[52] Hess JR, et al. Influence of aging on regeneration in end-to-side neurorrhaphy. Annals of Plastic Surgery. 2006;**57**(2):217–222

[53] Aichaoui F, Mertens P, Sindou M. Dorsal root entry zone lesioning for pain after brachial plexus avulsion: Results with special emphasis on differential effects on the paroxysmal versus the continuous components. A prospective study in a 29-patient consecutive series. PAIN®. 2011;**152**(8):1923–1930

[54] Bertelli JA, Ghizoni MF. Concepts of nerve regeneration and repair applied to brachial plexus reconstruction. Microsurgery. 2006;**26**(4):230–244

[55] Haninec P, et al. Usefulness of screening tools in the evaluation of long-term effectiveness of DREZ lesioning in the treatment of neuropathic pain after brachial plexus injury. BMC Neurology. 2014:**14**(1):225

[56] Nashold Jr BS, Ostdahl RH. Dorsal root entry zone lesions for pain relief. Journal of Neurosurgery. 1979;**51**(1):59–69

[57] Sindou MP, et al. Microsurgical lesioning in the dorsal root entry zone for pain due to brachial plexus avulsion: A prospective series of 55 patients. Journal of Neurosurgery. 2005;**102**(6):1018–1028

[58] Tomás R, Haninec P. Dorsal root entry zone (DREZ) localization using direct spinal cord stimulation can improve results of the DREZ thermocoagulation procedure for intractable pain relief. Pain. 2005;**116**(1–2):159–163

[59] Chen H, Tu Y. Long term follow-up results of dorsal root entry zone lesions for intractable pain after brachial plexus avulsion injuries. Acta Neurochir Suppl. 2006;99:73–75.

[60] Samii M, et al. Treatment of refractory pain after brachial plexus avulsion with dorsal root entry zone lesions. Neurosurgery. 2001;**48**(6):1269–1277

[61] Aly MM, et al. Importance of distinction between paroxysmal and continuous patterns of pain during evaluation of pain after brachial plexus injury. Acta Neurochirurgica. 2011;**153**(2):437–438

Section 2

Biomaterials, Neuroprotective Factors and Cell-Based Therapies for Peripheral Nerve Regeneration

Chapter 5

The Role of Nucleotides in Glial Cells during Peripheral Nerve Trauma and Compressive Disorders

Marina Manhães, Marcelo Cesar, Rayssa Justo,

Mauro Geller, Mendel Suchmacher and Rafael Cisne

Additional information is available at the end of the chapter

http://dx.doi.org/10.5772/68068

Abstract

Studies have shown that the administration of drugs containing pyrimidine nucleotides, such as uridine triphosphate (UTP) and cytidine monophosphate (CMP), has been effective in pain-intensity reductions in patients with painful conditions as diabetic neuropathy, back pain, and cervical and trauma-compressive changes. The combination of pyrimidine nucleotides UTP and CMP is part of a peripheral neuro-regenerative process. Its pharmacological properties are stimulation of nerve cells proteins synthesis, nerve cell membranes synthesis, myelin sheaths synthesis, and neurite sprouting through P2Y receptors activation. Herein, chapter will be discussed the combination of UTP and CMP, and in some cases, the inclusion of cobalamin (B12 vitamin) that appears to have analgesic effects in neuropathic pain secondary to spine structural disorders assigned to a complex pharmacodynamic. The mechanisms involved can be both indirect (protein synthesis in nerve cells, myelin synthesis, synthesis of MBP, etc.) and direct (P2Y receptor stimulation).

Keywords: nerve injury, nucleotides, peripheral regeneration, purinergic receptors, Schwann cells

1. Introduction

Neuropathic pain is defined as a pain caused by primary lesion or damage to the central or peripheral nervous system and is an issue that has not been thoroughly studied or resolved. Damage may result of compression, cutting, ischemic or metabolic disorders, cellular infiltration, or a combination of these factors [1]. About 50–90% of adults under 45 years, at some point of their lives, have a spine pain experience, especially in the lower back, being the main cause of disability [2]. Studies have shown that the administration of drugs containing pyrimidine nucleotides, such as uridine triphosphate (UTP) (**Figure 1A**) and cytidine monophosphate (CMP) (**Figure 1B**), has been effective reductions in pain intensity that have been reported in patients with painful conditions such as diabetic neuropathy, back pain, cervical pain, and trauma-compressive disorders [3–6]. The pyrimidine nucleotides UTP and CMP are part of a peripheral neuro-regenerative combination. Its pharmacological properties are stimulation of nerve cell synthesis of proteins, synthesis of nerve cell membranes, synthesis of myelin sheaths, and neurite sprouting through P2Y receptors stimulation [7]. Regarding analgesic capacity itself, pharmacological properties of two pyrimidinic nucleotides were experimentally demonstrated by Okada et al. (2010), which concluded that the activation of UTP-sensitive P2Y2 and/or P2Y4 receptors produces inhibitory effects on spinal pain transmission [8].

To better understand the role of nucleotides on peripheral nervous disorders, first, we need to get a brief review on peripheral nervous morphology as well as have the regeneration steps highlighted. The aim of this chapter is to clarify all steps and functions of those components in regeneration, focused on the relationship among nucleotides and glial cells.

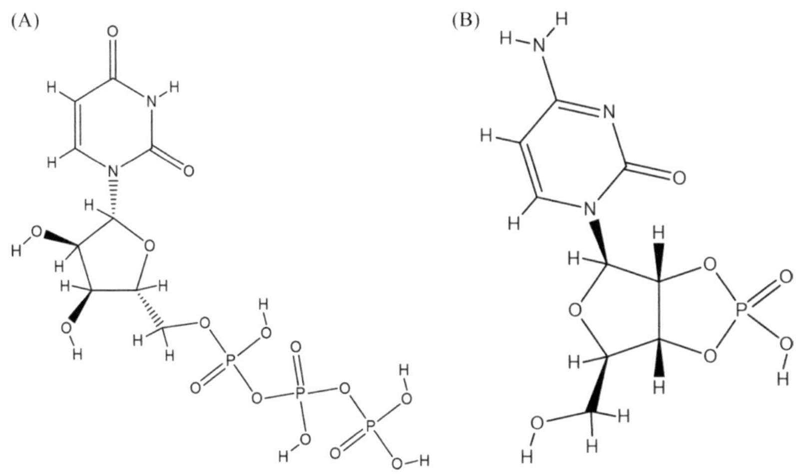

Figure 1. A—molecular structure of uridine triphosphate; **B**—molecular structure of cytidine monophosphate (CMP).

2. Peripheral nervous system

The peripheral nervous system (PNS) consists of (1) peripheral nerves, composed of the set of nerve fibers joined by connective tissue and (2) their motor and sensorial endings. In addition, nerves can be divided as their innervation—cranial or spinal—and as the types of fibers that compose them—sensorial, motor, or mixed [9].

The nervous tissue mainly consists of neurons and neuroglia, which helps in neuronal or defense activity, aiding in the support and protection of neurons. Each neuron has a cell body (soma) from which the axon radiates the nerve impulses to its synaptic terminal, and the dendrites, which receive and transmit synaptic information in the body of nerve cells [10]. In addition, some axons are surrounded by a myelin sheath, which in the central nervous system (CNS), is produced by oligodendrocytes and, in the PNS, by Schwann cells (SCs) [11]. In this way, the fibers are capable of conducting the electrical impulse, being called afferents when conducting to the CNS, or efferent when conduction starts from the CNS to the target organs [10].

3. Myelination

The myelinated fibers of the PNS are composed of a single axon, which is individually wrapped by a single SC [11, 12]. The membrane of SC surrounds the fiber to form a multilaminated myelin sheath [9] (**Figure 2**), isolating the axon and helping in the saltatory conduction of electrical signals [11].

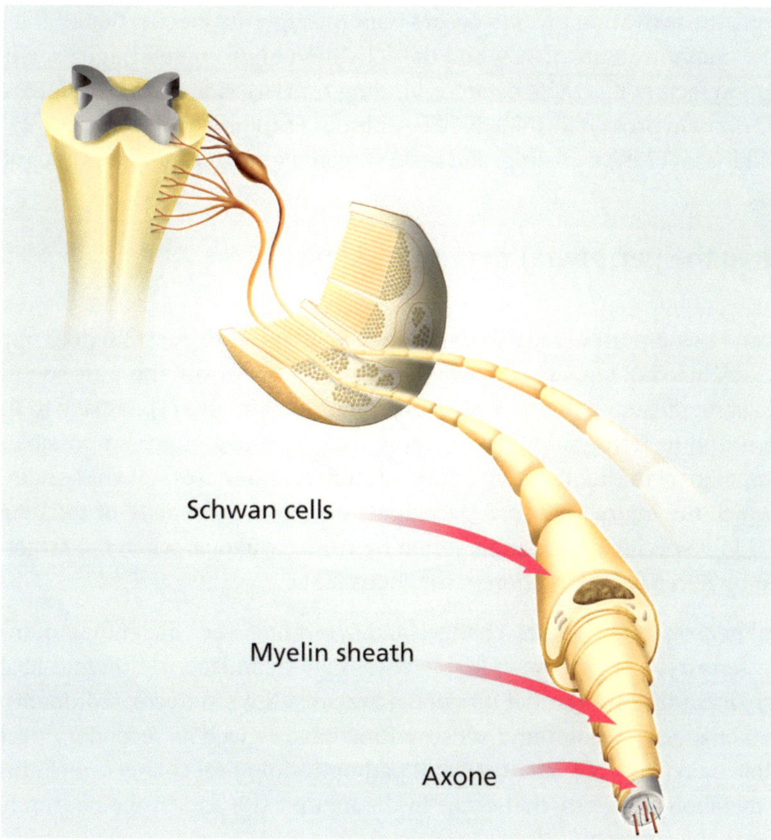

Figure 2. Diagram representing the origin of the neuroglial structural components.

Throughout the maturation process, all immature SCs have the same potential for development. When they are associated with axons of greater caliber (above 1 μm), they become myelinating SCs, and if they are associated with axons of small diameter, they become mature nonmyelinating SCs [13]. In this process, there is a fundamental participation of neurotrophin (NT), as nerve growth factor (NGF), which serves as a signaling for tyrosine kinase family receptors (TrkA) on axon, promoting an axonal diameter growth. Thus, NGF indirectly participates in myelinization of the axon. Besides that, it was observed that, during myelination, some cell adhesion molecules are downregulated, such as L1 and polysialylated neural cell adhesion molecule (NCAM), being expressed only in nonmyelinated axons [14].

The myelination done by a SC after its differentiation is closely linked both to its ability to synthesize a basal lamina and to its deposition. Furthermore, the presence of the axon and its intimate relationship with SC is extremely important, since it is in the axon that genes will be expressed then will maintain the myelinizing phenotype throughout the whole process [15], being the neurofascin gene an example, which plays an important role in the more advanced stages of myelination [14].

The axon also plays a key role in the activation through neuregulin of the ErbB2 and ErbB3 receptors of SCs, which are responsible for signaling the onset of the myelination process [16]. Therefore, the activation process occurs both through the inactivation of the signals that determine the immature state of SCs, and the activation of pro-myelin signals, which involve the transcription factors KROX20, octamer-binding transcription factor 6 (OCT6) and brain 2 class III POU domain protein (BRN2), NGFI-A-binding proteins 1 and 2 (NAB1/2), phosphatidylinositol 3-kinase (PI3K) signaling, and v-ski sarcoma viral oncogene homologue (SKI) [13].

4. Injuries to the peripheral nervous system

PNS axons can regenerate and recover their function after injury, fact that does not occur with CNS axons, which do not regenerate spontaneously [17]. However, there are some factors that contribute to an inefficient functional recovery, among which are (1) damage to the cell body of the neuron due to retrograde degeneration, making regeneration impossible; (2) nonviability of axonal growth due to nerve injury or subjacent diseases; (3) changes in the central circuits in which the injured neurons participate due to the plasticity of the neural connections; and (4) low specificity of reinnervation by the new axons, when the target organs are reinnervated by nerve fibers of different functions [18].

In peripheral neuropathic diseases, changes and symptoms vary depending of injured nerve type (motor, sensory, or autonomic) [19]. Hence, injury can lead to different levels of nerve fiber damage, including substantial functional loss, resulting in decreased quality of life due to permanent changes in motor and sensory functions, as well as secondary problems such as neuropathic pain [18]. As a result, several pathophysiological changes, including morphological and metabolic changes, that occur in the injured site and in the neuron body, in the proximal and distal segments [9].

These injuries are common, and their repair is still a problem in microsurgery. One method widely accepted by surgeons is to solve the problem with an autologous donor nerve, which is linked to some disadvantages, such as an extra incision to removal of a healthy sensory nerve, resulting in a sensory deficit [10]. On the other hand, in cases of chronic axotomy, the number of SCs in the distal stump decreases drastically, which makes the regeneration of axons difficult [20].

In this way, there are currently no repair techniques that ensure the recovery of normal sensory and motor functions after severe traumatic nervousness. Therefore, new therapeutic strategies are needed to potentiate axonal regeneration, promote selective reinnervation of the target, and modulate the central reorganization [18].

5. Wallerian degeneration

After nerve injury, the proximal fibers of the trauma are disconnected from the body of the neuron, resulting in a loss of muscles innervation, which leads to a total or partial loss of the motor, sensorial, and autonomic functions [18, 21]. In this way, a series of cellular alterations is initiated in the distal segment of the injured nerve, triggering the process of Wallerian degeneration, in which fragmentation and disintegration of the axons occur [17]. This disintegration is the result of a significant increase of Ca^{2+} in axoplasm, which is normally maintained at low concentrations in a healthy axon. On this way, Ca^{2+} sensitive protease (calpain) is activated, thereby degrading the axon cytoskeleton [22–24].

Wallerian degeneration also leads to removal and recycling of fragments derived from myelin rupture. For this, there is a recruitment of (1) macrophages, due to an increase in the permeability of blood-nerve-barrier (BNB) [24], contributing to removal of debris, phagocytizing them; and (2) SCs, which are dedifferentiated, divide and proliferate, also assisting in this removal and regulating factors that regulate Wallerian degeneration and nerve regeneration [12].

Besides that, several molecular changes are observed in the distal stump of the injured nerve, such as: (1) elevation of NGF messenger ribonucleic acid (mRNA) concentration, related to macrophage migration to the site and increased concentration of interleukin; (2) elevation of brain-derived neurotrophic factor (BDNF) mRNA concentration; (3) downregulation of NT-3 mRNA after nerve injury; (4) NT-4/5 mRNA decreases in the first hours after trauma but increases significantly after 2 weeks; (5) the expression of the transmembrane receptor for neurotrophic factors, p75NGFR, increases both in the distal stump and in the repair sites; (6) the expression of members of the tyrosine kinase family: trkA receptor is not detected, whereas the trkB and trkC levels in the SCs increase; (7) ciliary neutrophic factor (CNTF) mRNA decreases dramatically; and (8) rapid upregulation of glial cell line-derived neutrophic factor (GDNF) mRNA expression in SCs [25].

6. Peripheral nervous system regeneration

The main function of axonal regeneration is to replace the distal segment of the nerve that was lost during degeneration, allowing the reinnervation of peripheral segments and the restitution of their functions. Therefore, injured axons of peripheral nervous system are able to regenerate and reinnerve their target organs [21]. In view of this, while the degeneration process is happening in distal stump of the axon, the proximal stump regeneration begins, which occurs through the retrograde reaction that leads to metabolic changes [21, 26].

Moreover, axonic and myelinic debris were previously removed from the distal part of the injured site during the process of Wallerian degeneration by macrophages and CSs [12, 17]. The relationship between axons and SCs is intense and essential for regeneration process [27], since it is necessary a permissive environment for it. This is provided by the set of (1) extracellular matrix, (2) extracellular matrix proteins (ECMs) or neurostimulatory peptides (LN-1 or fragments of LN-1), (3) neutrophic factors, and (4) the SCs themselves [28].

Furthermore, SCs lose their myelinizing phenotype, leaded by a decrease in type III neuregulin 1. They become dedifferentiated and increase the expression of the growth factor-promoting genes [22], which aid the expansion of newly formed growth cones on the regenerating fibers. In addition, they regulate extracellular matrix molecules [20]. Thus, to aid the expansion of the axon, SCs increase their synthesis of adhesion molecules (CAMs), such as N-CAM, Ng-CAM/L1, N-cadherin, and L2/HNK-1; secrete ECM proteins, such as laminin (LN), fibronectin (FN), heparan sulfate proteoglycans (HSP), and tenascin in the basal membrane; secrete several neurotrophic factors, such as NGF and BDNF, to attract the fibers during their regeneration, being captured in the growth cones, incorporated into axon, and transported to the body of the neuron; and, together with macrophages, express anti-inflammatory cytokines such as interleukin (IL)-10 (NGEOW), which inhibit the inflammatory process initiated in Wallerian degeneration [25, 27].

Then, in the beginning of regeneration, it is possible to observe an axonal shoot appearing in the distal stump, while the surface of the SCs guides the growth cone, allowing the beginning of myelination [10]. This directional guidance track that provides way for axon growth is called Büngner band, which is made up by SCs [24], by the basal membrane where the SCs are situated [19], as well as by connective tissue. If the distance to be covered by the new axon segment is short, there may be a reinnervation in a healthy muscle. However, if reinnervation is delayed, SCs degenerate and no longer promote axon growth. Thus, in addition to atrophy in target muscle, the receptivity to synapse formation is lost [22].

In view of the active participation of SCs in the regeneration of peripheral nerves, the use of these cells has enabled the development of new strategies for the treatment of peripheral nervous disorders [29], including demyelinating diseases and spinal cord injuries [11].

7. Nucleotids as elements with therapeutical properties

It is known that extracellular nucleotides are fundamental in the regulation of several cellular and pathological mechanisms, being important in the control of homeostasis [30–32]. The regulation of the increase of other substances in cells, glucose and urea metabolism, and participation in inflammatory response processes are among these mechanisms [33, 34].

Nucleotides are monomeric structural units composed by a sugar moiety, attached to one or more phosphate groups, and a nitrogenous base, which may be cytosine, adenine, guanine, thymine, or uracil [35]. They are present inside cells playing a key role in several processes, such as the regulation of programmed cell death, energy generation, and cellular signaling [36].

The intracellular or physiological function performed by nucleotides is related to the type of receptor which this binds [37]. These receptors, known as purinoreceptors, are divided into two types: P1, which are adenine selective receptors, and P2, which are subdivided into P2X receptors, formed by ionotropic receptors of adenosine triphosphate (ATP), and P2Y coupled to G proteins, selective for nucleotides containing adenine and/or uracil [38].

This signaling modulates processes such as endocrine and exocrine secretion, platelet aggregation, cell proliferation, differentiation, bone resorption, inflammation, and healing [36]. In addition, P2Y receptors are related to cell survival or death mechanisms in order to promote tissue healing and regeneration, an important process in pathological conditions [39].

Several types of nucleotides—such as ATP, UTP and adenosine—act in the nervous system as signaling molecules in innumerable processes, such as neurogenesis, migration, neuron differentiation, apoptosis, and glial cell proliferation [40]. They may play a specific role, assisting in the development of the nervous system and its regeneration, in addition to participating in synaptic transmission and neuromodulation [41, 42].

Both the ATP-1 and UTP-2 nucleotides are mostly intracellular. However, both can be secreted into the extracellular medium by various mechanisms. One of them is the cellular damage, which leads to the release of nucleotides by necrotic or apoptotic cells, thus constituting a danger signal. Other mechanisms are exocytosis and transport by vesicles and membrane channels [43].

8. Nucleotids and cobalamin and their application for regeneration

The presence of extracellular nucleotides in the nervous system as signaling and regulatory molecules in several processes has been recognized for presenting neuromodulatory function involved in several stages of metabolism [44] and because they are potent microglial stimulators in both normal and pathophysiological pathways [45].

P2Y receptor ligands have been shown to be positively regulated in spinal microglial cells following damage in peripheral innervation, contributing for example by aiding the treatment of neuropathic pain and stimulating the release of neurotrophic factor from the brain [46]. In addition, extracellular nucleotides are capable of interacting with proximal cells, inducing cell differentiation and neurite outgrowth in glial cells [39, 47, 48]. Thus, they are molecules that, when induced, are effective in the treatment of several peripheral neurological syndromes, such as peripheral neuropathy [49, 50].

Another important role that nucleotides play is in the mechanism of macrophages recruitment as well as in the production of interleukins—such as IL-6, IL-9, and IL-13—via activation of P2Y and mRNA receptors [51–53]. The recruitment of macrophages to the injured site is essential for the regeneration of nervous tissue, since it promotes a rapid production of myelin in the PNS, as well as formation of myelin associated with glycoproteins and, therefore, facilitates nerve regeneration [54].

The interleukins mentioned above are important mediators of nerve regeneration, which act via interleukin receptors [55]. Studies show that IL-6 is not detected in intact nerves; however, in injured nerves, it is increased and it is regulated by neurotrophic factors, which are released by SCs [56].

Drugs containing nucleotides are prescribed, for example, to patients with neuromuscular diseases and diabetic polyneuropathy, since their clinical efficacy has already been studied, and *in vivo* tests have demonstrated their role in accelerating the regeneration of nerves and muscles after the sciatic lesion [29]. Mechanisms of tissue restoration have a vital importance for regeneration of the PNS, and nucleotides can be used as treatment for these lesions, since they play an important role in nerve regeneration [57].

In vitro studies show that UTP has an important costimulatory role in the wound healing process, activating, and modulating growth factors, which confirms the role of extracellular nucleotides in the process of tissue regeneration [58]. UTP, through the activation of P2Y purinergic receptors, induces in the SCs an N-cadherin expression increase which is closely related to growth and orientation of axons, besides having an important role in cell adhesion and myelination [59].

Derivatives of cytidine have been shown to be beneficial against various pathologies of the central nervous system, as well as neurodegenerative diseases. It is able to promote the regeneration of nerves in the peripheral nervous tissue and promotes the functional recovery of these nerves. Preclinical studies have shown that it promotes nerve regeneration in murine models. In addition, cytidine administered alone or in combination has an effect on peripheral nerve regeneration in rats, which compounds are believed to have the same function as cytidine in the regeneration of peripheral nerves [60, 61].

Evidences have shown that the combination of CMP and UTP has a positive effect on tissue regeneration [29, 62], as well as a meta-analysis study showed that P2Y receptor ligands are a promising therapeutic strategy for the treatment of neuropathic pain in murine models [63]. A clinical study of 26 patients with optic neuropathy–administered cytidine diphosphate (CDP)-choline for about 6 months showed that CDP-choline is effective in regenerating optic nerves in these patients [64].

Nunes et al. compared the efficacy of uridine and cytidine nucleotides associated or not with hydroxycobalamin (**Figure 3**) in the treatment of signs and symptoms of anemia. They observed that the group treated with the three elements achieved better efficacy—corresponding to an improvement in laboratory assessments, weight gain, and decreased pain—than the group treated only with nucleotides [65]. Another study evaluated the use of the three therapeutic elements in the treatment of patients with alcoholic polyneuropathy, and it was observed that their use was safe and effective, with decreased pain and improved motor coordination [66].

Further study carried out by Negrão et al. tested the use of uridine nucleotides associated with vitamin B12 and folic acid to assess the clinical improvement of patients with peripheral neuropathy associated with neuropathic pain. They observed a significant improvement in pain intensity, number of affected areas, and pain irradiation, suggesting a possible reduction in the use of nonsteroidal anti-inflammatory drugs (NSAIDs) [49].

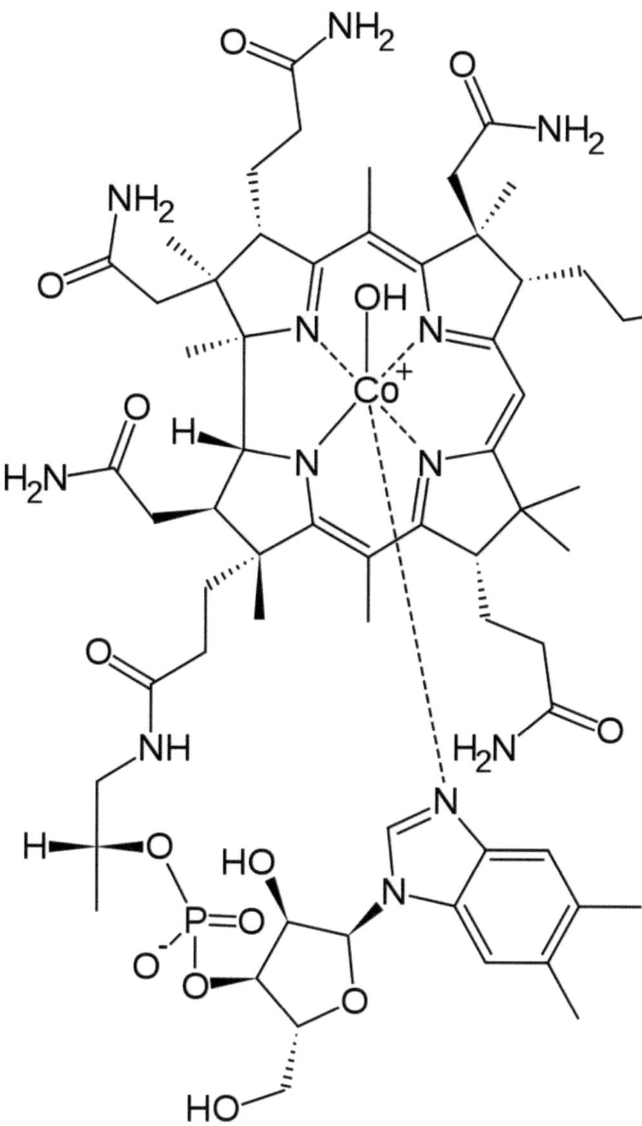

Figure 3. Molecular structure of cobalamin, Vitamin B12, whose central cobalt atom is attached to a hydroxyl radical.

Several meta-analysis studies have shown that there is a vitamin B12 deficiency in peripheral neuropathies due to type II diabetes, and that the administration of vitamin B12 in these patients is efficient as a treatment of neuropathy and neuropathic pain and may even be administered as prophylactic supplementation in this population [67–69].

Neurological disorders related to exposure of nitrous oxide anesthesia have been reported and are linked to toxicity in the spinal cord, since this substance causes irreversible oxidation of the cobalt ion present in the cobalamin structure [70]. In such cases, homocysteine methylation for S-adenosyl-methionine (SAM) formation is defective, leading to the formation of unstable myelin basic proteins [71]. Also, clinical study has shown that parenteral administration of vitamin B12 in a series of cases with different neurological abnormalities, where patients had vitamin B12 deficiency were effective for the treatment of peripheral neurological damage [72–75].

Vitamin B12 plays an important role in DNA synthesis and neurological functions, and its deficiency induces a failure of the methylation of basic myelin proteins and may be the cause of myeloneuropathy or peripheral neuropathy [76]. Weir and Scott showed that B12 deficiency is very common in the elderly and is important in the brain where SAM synthesis occurs [77]. In addition, other pathophysiological conditions such as survivors of acute lymphoblastic leukemia during childhood, patients with rare Foster Kennedy syndrome, or patients with nitric oxide toxicity, may present neuropathy due to vitamin B12 deficiency, and in these cases, the administration of it is used as treatment [76–79].

Futhermore, it is known that vitamin B12 deficiency causes neurological changes that form a classic clinical picture of subacute degeneration of the dorsal and lateral vertebral column as a consequence of changes in myelin formation [80, 81] and in that cases, the standard treatment is the administration of cobalamin [82]. Besides, B vitamins have an analgesic effect in painful neuropathic and nociceptive syndromes [83].

9. Molecular perspectives

Endogenous substances when administered exogenously tend to be processed as elements belonging to normal physiology, in which homeostatic mechanisms act to bring them back to their normal levels [84, 85]. The control of blood levels of nucleosides is exerted by the balance between three different metabolic pathways: (1)—hepatic *de novo* synthesis, (2)—salvage pathway, (3)—hepatic degradation [86–88]. Both uridine and cytidine pass into the nervous system from the choroid plexus and the blood-brain barrier, through nucleoside transport systems [89].

These systems are divided into low-affinity equilibrium transport system (SLC29 family) and high-affinity concentration transport system, which is sodium-dependent, substrate-selective and unidirectional (SLC28 family) [90]. Both the transport of blood to the cerebral extracellular fluid and the extracellular cerebral fluid to the neural cells are mediated by these transporters [91, 92].

Oral administration of cytidine to humans rapidly elevates uridine serum levels because of the conversion of part of it into uridine [93]. Therefore, even if administration of exogenous cytidine leads to increased levels of its nucleoside in neural cells, since the uridine is the main precursor for CTP used in the synthesis of brain phosphatides [89] (see **Figure 4**).

The biosynthesis of phosphatidylcholine, the most abundant phospholipid in the brain and phosphatidylethanolamine proceeds through activation of the amine moiety (namely choline or ethanolamine) by coupling to CDP prior to its addition to the diacylglycerol, leading to the production of CDP-choline or CDP ethanolamine and inorganic phosphate [94]. Both cytidine and uridine are able to increase neuronal membrane synthesis through increasing levels of CTP [95–97].

http://dx.doi.org/10.5772/68068

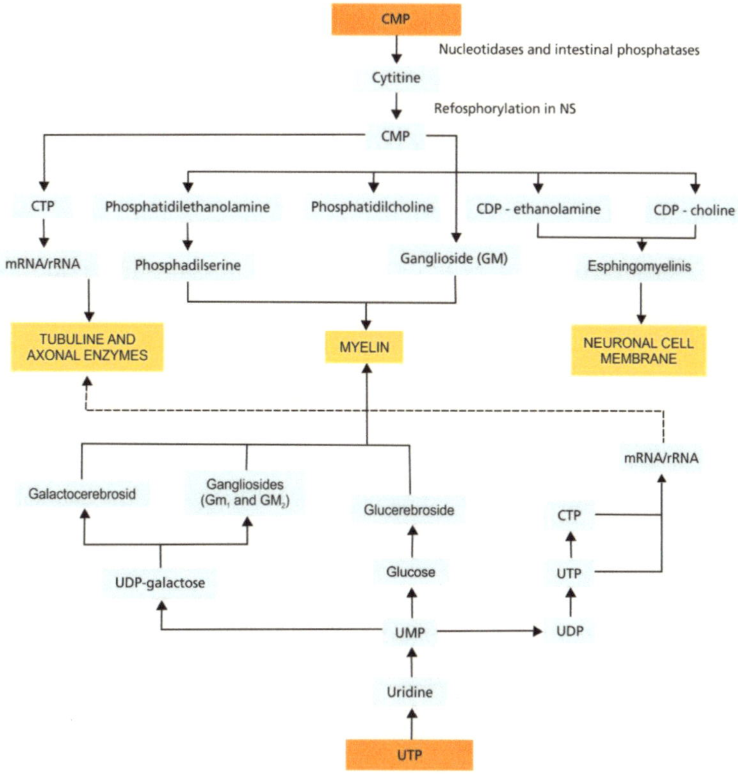

Figure 4. Overview of metabolic pathways (protein synthesis, myelin sheath synthesis, and neuronal cell membrane synthesis), to which exogenous nucleotides CMP and UTP integrate. Abbreviations: **CDP:** cytidine diphosphate; **CMP:** cytidine monophosphate; **CTP:** cytidine triphosphate; **mRNA:** messenger ribonucleic acid; **NS:** nervous system; **rRNA:** ribosomal ribonucleic acid; **UDP:** uridine diphosphate; **UMP:** uridine monophosphate; **UTP:** uridine triphosphate.

The circulating pyrimidines, in addition to being incorporated into nucleic acids, may serve as substrates for the salvage route of pyrimidine nucleotide synthesis, as precursors of cytidine triphosphate (CTP) [98] and as precursors for uridine diphosphate (UDP) and uridine triphosphate (UTP), which activate the brain's P2Y receptors [99].

Cytidine nucleotides are extremely important for the replacement of phospholipids that serve as substrates for cell membrane synthesis in the nervous system, such as phosphatidylinositol, phosphatidylcholine and phosphatidylethanolamine [89, 100]. In addition, they are also involved in the modulation of pain transmission by the activation of P2Y receptors [101].

Uridine nucleotides activate specific P2Y receptor subtypes in humans [102, 103], acting as cell-to-cell signaling in the nervous system [104, 105]. They are dependent on the activity of the axonal signals in neighboring oligodendrocytes and their structure consists of seven transmembrane domains, with the N-terminal domain in the extracellular space and the C-terminal domain in the cytoplasm [19, 106]. In addition, activation of P2Y receptors is usually associated with the stimulation of various mitogen-activated protein kinases (MAPKs), mainly the extracellular signal-regulated protein kinase 1/2 [38].

The activation of purinergic receptors in axons and SCs in regeneration processes is vital, since their inhibition leads to improper regeneration of the nerve [107]. Physiologically, extracellular UTP is capable of causing secretion in calcium chloride epithelial cells and in glial cells of catecholamines. In the SCs, UTP treatment contributes to the increase of the excitatory communication between axons and these cells through the secretion of ATP [19] and increased N-cadherin expression, an adhesion protein that could reanalyze the early contacts between cells and axons to accelerate myelination and axonal regeneration [59].

Subtypes of UTP-activated receptors P2Y2 and P2Y4 in humans are coupled to G_q protein [102] and are mainly involved in long-term effects, such as differentiation, neurite outgrowth, and cell survival or death [108, 109]. These receptors are normally activated during pathological conditions and participate in inflammatory processes of the nervous system [102], in which they trigger and sustain reactive astrogliosis, the reaction to brain trauma [104, 105], characterized by cellular proliferation and neural circuit remodeling [110].

The P2Y2 and P2Y4 receptors activate phospholipase C, increasing the cytosolic Ca^{2+} concentration from the intracellular reserves and the activation of protein kinase C in response to the production of inositol 1,4,5-trisphosphate and diacylglycerol, respectively [111]. Generally, P2Y receptors that increase intracellular Ca^{2+} concentration induce the tricarboxylic acid cycle and increase ATP production, which promotes the maintenance of ion homeostasis and antioxidant defense [109].

P2Y2 receptors are expressed by neurons, astrocytes, and microglia and regulate actin polymerization and cytoskeletal rearrangements through the Rac/Rho pathways [112], as well as the P2Y4 receptors [113]. Its activation confers neuroprotection in several ways: the promotion of neurite outgrowth, increased cell motility, nonamyloidogenic processing of the amyloid precursor protein, and increased phagocytosis and degradation of the amyloid-beta peptide [102, 111]. Moreover, studies have shown that microglia respond rapidly to nerve lesions by migrating to the spinal projection territories of the central terminals of injured primary afferents, with subsequent proliferation, activation of p38 MAPK and ERK1/2, and production of proinflammatory cytokines and chemokines [114].

UTP also participates in neuromodulation. The modulation exerted by activation of P2Y4 receptors is linked to the positive influence on excitatory transmission mediated by postsynaptic N-methyl-D-aspartate (NMDA) and α-amino-3-hydroxy-5-methyl-4-isoxazolepropionic acid (AMPA) receptors, through increased glutamate release [111], while P2Y2 receptors modulation is linked to increased currents through the Ca^{2+} permeable transient receptors potential vanoloyide 1 (TRPV1) in the PNS [115]. This high concentration of intracellular Ca^{2+} can participate in responses through annexins—responsible for signal transduction, trafficking, and vesicle aggregation and membrane organization—that bind negatively charged phospholipids in a Ca^{2+}-dependent way [116].

In addition to UTP, other uridine derivatives also activate P2Y receptors, such as UDP, which activate the subfamily of P2Y6 and P2Y14 receptors [117, 118]. However, while P2Y6 receptors are coupled to G_q/11 protein, P2Y14 receptors are coupled to G protein, and the increase in their expression is also related to the occurrence of peripheral nerve lesions, being regulated by p38 MAPK [114]. The activation of the P2Y6 receptors in microglia cells causes a rapid change in their morphology, with phagocytosis of damaged neurons being increased, through the reorganization of actin by a pathway mediated by the activation of protein kinase C (PKC) and PCL linked to the increase of intracellular Ca^{2+} [94].

Cobalamin and its analogs act on the nervous system promoting neurite outgrowth and neuronal survival. It plays the role of coenzyme in the methylation of homocysteine by methionine synthase to form methionine, in isomerization of 1-methylmalonyl-CoA in succinyl-CoA catalyzed by 1-methylmalonyl-coenzyme A mutase [119, 120], and in the activation of Erk1/2 and Akt [8].

In addition to the formation of methionine from homocysteine, methionine synthase is required for the synthesis of S-Adenosyl-methionine [121, 122]. SAM is a key metabolite in amino acid transmethylation responsible for the biological methylations that modify nucleic acids, fatty acids, porphyrins, phospholipids, polysaccharides, biogenic amines, and proteins, such as myelin basic protein (MBP), one of the proteins responsible for the compaction of the cytoplasmic surfaces of the myelin sheath [123, 124].

Vitamin B12, as the effector of methionine synthase, plays a key role in ensuring the integrity and stability of the myelin basic protein, since it depends on the methylation of one of its amino acids. A deficiency in this methylation can lead to poor protein formation and instability [119]. In addition, methionine also facilitates the formation of formyl tetrahydrofolate (formyl THF) and tetrahydrofolate (THF), which are involved in the synthesis of purines [125].

Under normal conditions, when the folding and/or trafficking of a polypeptide fails, the protein is targeted for degradation by the ubiquitin-proteasome system. However, when there is a pathologic condition, there is interruption of the balance between the synthesis/folding and degradation pathways, and the accumulation and aggregation of proteins are favored, which are a characteristic of several neurodegenerative diseases [126].

10. Discussion

Some clinical studies using nucleotides and vitamin B12 for the treatment of diseases of the peripheral nervous system have been carried out, proving that CMP, UTP, and hydroxycobalamin are effective and can be used safely for this purpose. The studies cited below are summarized in **Table 1**.

Lauretti et al. evaluated the efficacy of oral administration of the cytidine-uridine-hydroxycobalamin complex in the treatment of chronic neuropathic lower back pain. The study evaluated 48 adult patients, aged 21–80 years, with a history of pain after 6 months, whose previous traditional treatments were ineffective. During the course of treatment, patients were given oral fluoxetine (20 mg/day) daily and were divided into two groups: the control group, which received a combination of 40-mg lidocaine, 30-mg clonidine, and 10-mg dexamethasone, diluted with physiological solution; and study group where a tablet containing

Authors	Type of study	Sample	Objectives/methods	Results/conclusion
Parisi et al. [64]	Clinical study of the effects of CDP-choline on patients with optic neuropathy	26 patients were treated with the disease and compared with 14 normal individuals	Patients were treated with oral CDP-choline for two 60-day periods and one later to complete 360 days of study initiation. The results were evaluated by electrophysiological exams.	There was a significant improvement with the treatment of CDP-choline patients
Nunes et al. [66]	Clinical study of the use of CMP UTP and B12 to treat alcoholic polyneuropathy	120 patients between 18 and 65 years of age were evaluated	Patients with alcoholic neuropathy were treated with CMP, UTP and B12 intramuscularly in 6 days and orally for 30 days and then the effects were monitored	The combination of uridine, cytidine, and vitamin B12 was safe and effective in the treatment of patients with alcoholic neuropathy
Mibielli et al. [7]	The analgesic effects of the combination UTP, CMP and hydroxycobalamin were evaluated in a self-paired evolutionary model	17 men and 24 women were treated	Analysis of previously unpublished data from investigators files on VAS and PFQ pain scores of the group of patients treated with the combination of UTP, CMP and hydroxycobalamin	The combination of UTP, CMP and hydroxycobalamin seems to have analgesic properties in the medium term
Negrão et al. [49]	Clinical evaluation of patients with peripheral neuropathy and neuropathic pain	212 patients with a mean age of 59 (±14.4) years of age	Patients received daily treatment of uridine monophosphate + folic acid + vitamin B12 for 2 months in conjunction with anti-inflammatories and were evaluated using a pain-detection questionnaire	The combination of UMP + vitamin B12 + folic acid is effective against neuropathic pain associated with peripheral neuropathy. The use of anti-inflammatory decreased by more than 70%.
Negrão and Nunes [50]	Observational study of patients with neuropathy	48 patients were evaluated	Patients received daily treatment of uridine monophosphate + folic acid + vitamin B12 for 2 months in conjunction with analgesics and anti-inflammatories, and were evaluated using a pain-detection questionnaire	Uridine monophosphate + folic acid + vitamin B12 reduced total pain score, intensity and characterization of pain and associated symptoms and the use of analgesic and anti-inflammatory drugs reduced in 77.4%

Authors	Type of study	Sample	Objectives/methods	Results/conclusion
Solomon [68]	A clinical study that measured the amount of B12 in cancer patients associating neuropathy and neuropathic pain with B12 deficiency and consequent increase in malignancy	241 patients were evaluated	We evaluated the levels of B12 and malignancy in individuals with cancer during a period of 4 years in a cancer study center	B12 therapy plays an important role in the prevention of neuropathy and neuropathic pain that are generally encountered during cancer malignancy advances

Table 1. Summarized clinical studies using nucleotides and vitamin B12 for the treatment of diseases of the peripheral nervous system.

the cytidine-uridine-hydroxycobalamin complex was added and given orally every 12 hours. The results of the study indicated that the co-administration of the complex during treatment led to a decrease in the intensity of chronic neuropathic low back pain and a reduction in the consumption of rescue analgesics, improving and enhancing the quality of treatment in patients with neuropathic low back injuries [6].

Parisi and colleagues used CDP choline to treat optic neuropathy in a study with 26 sick patients in the test group and 14 healthy subjects in the control group. The treatment was done orally for two 60-day periods and a later period until it was completed 360 days after the start of the study. The results were evaluated through electrophysiological examinations, leading to a significant improvement in the patients. Thus, it has been found that CDP choline can be used to treat patients with optic neuropathy [64].

Goldberg and colleagues evaluated the use of a combination of uridine triphosphate (UTP), cytidine monophosphate (CMP), and hydroxocobalamin in a double-blind, randomized study in the treatment of neuralgia due to degenerative orthopedic alterations with neural compression. The patients were divided into two groups, being Group A: total daily dose of 9 mg UTP, 15 mg CMP, 6 mg hydroxycobalamin; and Group B: total daily dose of 6 mg hydroxocobalamin. At the end of the 30-day treatment period, there were reductions in the pain scale scores in both groups; however, there was a significantly larger reduction in the scores of the Group A patients. Based on these findings, the authors concluded that the combination of UTP, CMP, and vitamin B12 has a positive effect on pain and functionality improvement in the treatment of degenerative orthopedic alterations with neural compression [127].

Nunes and collaborators administered CMP, UTP, and hydroxycobalamin in patients with alcoholic polyneuropathy, a disorder in the peripheral nervous system involving motor, sensorial, and autonomic nerves. This study included 120 patients aged 28–65 years, who were treated with doses intramuscularly for 6 days and orally for 30 days. Afterward, the efficacy of the treatment was evaluated through sensorial motor tests as well as a visual evaluation of pain. With this, it was concluded that the treatment was effective and safe, reducing pain and improving the motor activity of the patients [66].

Negrão et al. performed an observational clinical study of 212 patients with peripheral neuropathy and neuropathic pain, treated orally for 2 months with capsules of uridine monophosphate (UMP), folic acid, and hydroxycobalamin. These patients had a mean age of 59 years. The results were evaluated using a questionnaire, where the patient reveals to the doctor the areas of the body that present pain. The result showed that the treatment was effective, and the statistical analysis showed that there was improvement not only of the overall picture but also factors such as intensity of pain and affected areas decreased with treatment. In addition, patients greatly reduced the adjunctive use of analgesic or anti-inflammatory drugs [49].

In another study, the same group of researchers treated 48 patients with neuropathy or neuropathic pain over a two-month period. Patient evaluations showed a reduction in intensity and areas affected by pain, showing the efficacy of the treatment and confirming the results of the previous study. As in the previous study, the treatment induced improvement of the patients allowing the reduction of analgesic or anti-inflammatory use by up to 70% [50].

Mibielli et al. conducted a clinical study to evaluate the analgesic effects of UTP, CMP, and hydroxycobalamin in the treatment of peripheral pain. A total of 17 men and 24 women with a mean age of 49 years were treated with oral administration of the compound. The evaluation of the results was performed with a pain questionnaire. The study showed that this compound presents analgesic and neuroregenerative properties in the medium term, as well as indicated subsequent randomized clinical trials to confirm the results [7].

The increase in the malignancy of some types of cancer is associated with the appearance of peripheral neuropathies. A recent study evaluated, over 2 years, levels of vitamin B12 in cancer patients at a cancer study center. From the analyses made, it was verified that vitamin B12 deficiency is associated with the appearance of neuropathies and peripheral pain. Therefore, treatment with vitamin B12 may prevent cancer patients from developing neuropathies or neuropathic pain [68].

11. Conclusion

New therapeutic strategies are needed to potentiate regeneration in nerve injuries, and the use of Schwann cells has enabled the development of new strategies for the treatment of peripheral nervous disorders. Numerous studies have been done with the aim of finding new targets and new drugs, and the use of uridine and cytidine nucleotides associated with hydroxocobalamin has proven to be very effective.

It can be assumed that the nucleotide supplementation of cytidine and uridine associated with vitamin B12 in situations of neural structural regeneration can increase its availability in SCs, aiding in neuro-regeneration. Therefore, it is a set of drugs that can be used safely in the treatment of neuropathies and other diseases associated with degeneration of the peripheral nervous system.

http://dx.doi.org/10.5772/68068

Abbreviations

AMPA	α-Amino-3-hydroxy-5-methyl-4-isoxazolepropionic acid
ATP	Adenosine triphosphate
BDNF	Brain-derived neurotrophic factor
BNB	Blood-nerve-barrier
CAM	Cell adhesion molecule
CDP	Cytidine diphosphate
CMP	Cytidine monophosphate
CNS	Central nervous system
CNTF	Ciliary neutrophic factor
CTP	Cytidine triphosphate
D	Day
DNA	Deoxyribonucleic acid
ECMs	Extracellular matrix
ERK	Extracellular signal-regulated kinase
FN	Fibronectin
GDNF	Glial cell line-derived neutrophic factor
IL	Interleukin
LN	Laminin
M	Month
MAPKs	Mitogen-activated protein kinase
MBP	Myelin basic protein
NGF	Nerve growth factor
NMDA	N-methyl-D-aspartate
NSAIDs	Nonsteroidal antiinflammatory drugs
NT	Neurotrophin
OCT6	Octamerbinding transcription factor 6
PI	Phosphatidylinositol
PKC	Protein kinase C
PLC	Phospholipase C
PNS	Peripheral nervous system
RNA	Ribonucleic acid
SAM	S-adenosyl-methionine

SCs	Schwann cells
THF	Tetrahydrofolate
Trk	Tyrosine kinase family
TRPV1	Transient receptor potential vanoloyide 1
UDP	Uridine diphosphate
UMP	Uridine monophosphate
UTP	Uridine triphosphate
V	Visit
VAS	Visual analogic scale

Author details

Marina Manhães[1], Marcelo Cesar[1], Rayssa Justo[1], Mauro Geller[2,], Mendel Suchmacher[3] and Rafael Cisne[1*]

*Address all correspondence to: rafael.cisne@gmail.com

1 Fluminense Federal University, Niterói, Brazil

2 Rio de Janeiro Federal University, Rio de Janeiro, Brazil

3 University Center Serra dos Órgãos, Teresópolis, Brazil

References

[1] Woolf CJ. Neuropathic pain. In: Pain. Philadelphia: Lippincott Williams Wilkins, 2004: 765–95.

[2] Silva GNS, Monteiro PB, Virgínio NA, Souto CGV, Oliveira MVAV. Systematization of nursing care for patients with Herniated Disc. Health Science Magazine New Hope. 2013;11(2):55–71.

[3] Serra JP, Veciana MG, Bordas LB. Therapeutic effects of an association of C.M.P., U.T.P., and vitamin B12 in fifty cases of peripheral neuropathies. New England Journal of Medicine (Sp Ed). Mayo 1972;VI(64):1–7.

[4] Gallai V, Mazzotta G, Montesi S, Sarchielli P, et al. Effects of uridine in the treatment of diabetic neuropathy: an electrophysiological study. Acta Neurologica Scandinavica. 1992;86:3–7.

[5] Muller D. Treatment of neuropathic pain syndrome. Results of an open study on the efficacy of a pyrimidine nucleotide preparation. Fortschritte der Medizin Originalien. 2002;120:131–3.

[6] Lauretti GC, Omals M, Pereira AC, et al. Clinical evaluation of the analgesic effect of the cytidine-uridine-hydroxocobalamin complex as a coadjuvant in the treatment of chronic neuropathic low back pain. Column. 2004;3(2):73–6.

[7] Mibielli MA, Nunes CP, Scussel AB, Suchmacher Neto M, Oliveira L, Geller M. Symptomatic improvement in an acute, non-traumatic spine pain model with a combination of uridine triphosphate, cytidine monophosphate, and hydroxocobalamin. Pain Studies and Treatment. 2014;v. 02, pp. 6–10.

[8] Okada K, Tanaka H, Temporin K, Okamoto M, Kuroda Y, Moritomo H, et al. Methylcobalamin increases Erk1/2 and Akt activities through the methylation cycle and promotes nerve regeneration in a rat sciatic nerve injury model. Experimental Neurology. 2010;222(2):191–203.

[9] Geuna S, Raimondo S, Ronchi G, Di Scipio F, Tos P, Czaja K, Fornaro M. Histology of the peripheral nerve and changes occurring during nerve regeneration. International Review of Neurobiology. 2009;87:27–46.

[10] Johnson EO, Zoubos AB, Soucacos PN. Regeneration and repair of peripheral nerves. Injury. 2005;36(4):S24–9.

[11] Bhatheja K, Field J. Schwann cells: origins and role in axonal maintenance and regeneration. The International Journal of Biochemistry & Cell Biology. 2006;38(12):1995–9.

[12] Campana WM. Schwann cells: activated peripheral glia and their role in neuropathic pain. Brain, Behavior, and Immunity. 2007;21(5):522–7.

[13] Jessen KR, Mirsky R. The origin and development of glial cells in peripheral nerves. Nature Reviews. Neuroscience. 2005;6(9):671–82.

[14] Sherman DL, Brophy PJ. Mechanisms of axon ensheathment and myelin growth. Nature Reviews. Neuroscience. 2005;6(9):683–90.

[15] Garbay B, Heapec AM, Sargueila F, Cassagne C. Myelin synthesis in the peripheral nervous system. Progress in Neurobiology. 2000;61(3):267–304.

[16] Pereira JA, Baumann R, Norrmén C, Somandin C, Miehe M, Jacob C, Lühmann T, Manten N, Meijer D, Suter U. Dicer in Schwann cells is required for myelination and axonal integrity. Journal of Neuroscience. 2010;30(19):6763–75.

[17] Huebner EA, Strittmatter SM. Axon regeneration in the peripheral and central nervous systems. In: Cell Biology of the Axon. 2009. Springer Berlin Heidelberg:305–60.

[18] Navarro X. Neural plasticity after nerve injury and regeneration. International Review of Neurobiology. 2009;87:483–505.

[19] Canales TM. Study of the drug CMP Forte and nucleotide UTP in Schwann cells [thesis]. Barcelona: International University of Catalonia; 2011.

[20] Jonsson S, Wiberg R, McGrath AM, Novikov LN, Wiberg M, Novikova LN, Kingham PJ. Effect of delayed peripheral nerve repair on nerve regeneration, Schwann cell function and target muscle recovery. PloS One. 2013;8(2):e56484.

[21] Allodi I, Udina E, Navarro X. Specificity of peripheral nerve regeneration: interactions at the axon level. Progress in Neurobiology. 2012;98(1):16–37.

[22] Scheib J, Höke A. Advances in peripheral nerve regeneration. Nature Reviews. Neurology. 2013;9(12):668–76. DOI: 10.1038/nrneurol.2013.227.

[23] Gaudet AD, Popovich PG, Ramer MS. Wallerian degeneration: gaining perspective on inflammatory events after peripheral nerve injury. Journal of Neuroinflammation. 2011;8(1):110.

[24] Ngeow WC. Scar less: a review of methods of scar reduction at sites of peripheral nerve repair. Oral Surgery, Oral Medicine, Oral Pathology, Oral Radiology, and Endodontology. 2010;109(3):357–66.

[25] Frostick, SP, Qi Yin, Kemp GJ. Schwann cells, neurotrophic factors, and peripheral nerve regeneration. Microsurgery. 1998;18(7):397–405.

[26] Zu-Lin C, Yu W-M, Strickland S. Peripheral regeneration. Annual Review of Neuroscience. 2007;30:209–33.

[27] Webber C, Zochodne D. The nerve regenerative microenvironment: early behavior and partnership of axons and Schwann cells. Experimental Neurology. 2010;223(1):51–9.

[28] Bellamkonda RV. Peripheral nerve regeneration: an opinion on channels, scaffolds and anisotropy. Biomaterials. 2006;27(19):3515–8.

[29] Martiáñez T, Carrascal M, Lamarca A, Segura M, Durany N, Masgrau R, Abian J, Gella A. UTP affects the Schwannoma cell line proteome through P2Y receptors leading to cytoskeletal reorganisation. Proteomics. 2012;12(1):145–56.

[30] Baer HP, Drummond GI. Physiological and regulatory functions of adenosine and adenine nucleotides. 1979. Raven Press.

[31] Rudolph FB. The biochemistry and physiology of nucleotides. The Journal of Nutrition. 1994;124(1 Suppl):124S–7.

[32] Corriden R, Insel PA. New insights regarding the regulation of chemotaxis by nucleotides, adenosine, and their receptors. Purinergic Signalling. 2012;8(3):587–98.

[33] Fausther M, Sévigny J. Nucleosides and extracellular nucleotides regulate hepatic functions through a complex system of membrane proteins. Biological Complications. 2011;334(2):100–17.

[34] Zimmermann H. Extracellular ATP and other nucleotides—ubiquitous triggers of intercellular messenger release. Purinergic Signalling. 2016;12(1):25–57.

[35] Florian J, Leszczyński J. Spontaneous DNA mutations induced by proton transfer in the guanine. Cytosine base pairs: an energetic perspective. Journal of the American Chemical Society. 1996;118(12):3010–7.

[36] Burnstock G. Pathophysiology and therapeutic potential of purinergic signaling. Pharmacological Reviews. 2006;58(1):58–86.

[37] Jacobson KA, Paoletta S, Katritch V, Wu B, Gao Z-G, Zhao Q, et al. Nucleotides acting at P2Y receptors: connecting structure and function. Molecular Pharmacology. 2015;88(2):220–30.

[38] Burnstock G. Purine and pyrimidine receptors. Cellular and Molecular Life Sciences. 2007;64(12):1471–83.

[39] Mamedova LK, Gao ZG, Jacobson KA. Regulation of death and survival in astrocytes by ADP activating P2Y 1 and P2Y 12 receptors. Biochemical Pharmacology. 2006;72(8):1031–41.

[40] Ulrich H, Abbracchio MP, Burnstock G. Extrinsic purinergic regulation of neural stem/progenitor cells: implications for CNS development and repair. Stem Cell Reviews and Reports. 2012;8(3):755–67.

[41] Neary JT, Zimmermann H. Trophic functions of nucleotides in the central nervous system. Trends in Neurosciences. 2009;32(4):189–98.

[42] Zimmermann H, Braun N. Extracellular metabolism of nucleotides in the nervous system. Journal of Autonomic Pharmacology. 1996;16(6):397–400.

[43] Jacobson KA, Boeynaems JM. P2Y nucleotide receptors: promise of therapeutic applications. Drug Discovery Today. 2010;15(13):570–8.

[44] Ipata PL. Origin, utilization, and recycling of nucleosides in the central nervous system. Advances in Physiology Education. 2011;35(4):342–6.

[45] Inoue K. The function of microglia through purinergic receptors: neuropathic pain and cytokine release. Pharmacology & Therapeutics. 2006;109(1):210–26.

[46] Tozaki-Saitoh H, Tsuda M, Miyata H, Ueda K, Kohsaka S, Inoue K. P2Y12 receptors in spinal microglia are required for neuropathic pain after peripheral nerve injury. Journal of Neuroscience. 2008;28(19):4949–56.

[47] Guarnieri S, Pilla R, Morabito C, Sacchetti S, Mancinelli R, Fanò G, Mariggiò MA. Extracellular guanosine and GTP promote expression of differentiation markers and induce S-phase cell-cycle arrest in human SH-SY5Y neuroblastoma cells. International Journal of Developmental Neuroscience. 2009;27(2):135–47.

[48] Menéndez-Méndez A, Díaz-Hernández JI, Miras-Portugal MT. The vesicular nucleotide transporter (VNUT) is involved in the extracellular ATP effect on neuronal differentiation. Purinergic Signalling. 2015;11(2):239–49.

[49] Negrão L, Almeida P, Alcino S, Duro H, Libório T, Melo SU, Figueira R, Gonçalves S, Neto PL. Effect of the combination of uridine nucleotides, folic acid and vitamin B12 on the clinical expression of peripheral neuropathies. Pain management. 2014;4(3):191–6.

[50] Negrão L, Nunes P. Uridine monophosphate, folic acid and vitamin B12 in patients with symptomatic peripheral entrapment neuropathies. Pain. 2016;6(1):25–9.

[51] Kobayashi D, Ohkubo S, Nakahata N. Contribution of extracellular signal-regulated kinase to UTP-induced interleukin-6 biosynthesis in HaCaT keratinocytes. Journal of Pharmacological Sciences. 2006;102(4):368–76.

[52] Endo Y, Isono K, Kondo M, Tamaoki J, Nagai A. Interleukin-9 and Interleukin-13 augment UTP-induced Cl ion transport via hCLCA1 expression in a human bronchial epithelial cell line. Clinical & Experimental Allergy. 2007;37(2):219–24.

[53] Zhang Z, Wang Z, Ren H, Yue M, Huang K, Gu H, Liu M, Du B, Qian M. P2Y6 agonist uridine 5′-diphosphate promotes host defense against bacterial infection via monocyte chemoattractant protein-1–mediated monocytes/macrophages recruitment. The Journal of Immunology. 2011;186(9):5376–87.

[54] Stoll G, Jander S, Myers RR. Degenerarion and regeneration os the peripheral nervous system: from Augustus Waller's observations to neuroinflammation. Journal of the Pheripheral Nervous System. 2002;7(1):13–27.

[55] Leibinger M, Müller A, Gobrecht P, Diekmann H, Andreadaki A, Fischer D. Interleukin-6 contributes to CNS axon regeneration upon inflammatory stimulation. Cell Death & Disease. 2013;4(4):e609.

[56] Hirota H, Kiyama H, Kishimoto T, Taga T. Accelerated nerve regeneration in mice by upregulated expression of interleukin (IL) 6 and IL-6 receptor after trauma. Journal of Experimental Medicine. 1996;183(6):2627–34.

[57] Rochkind S, Rousso M, Nissan M, Villarreal M, Barr-Nea L, Rees DG. Systemic effects of low-power laser irradiation on the peripheral and central nervous system, cutaneous wounds, and burns. Lasers in Surgery and Medicine. 1989;9(2):174–82.

[58] Boucher I, Kehasse A, Marcincin M, Rich C, Rahimi N, Trinkaus-Randall V. Distinct activation of epidermal growth factor receptor by UTP contributes to epithelial cell wound repair. The American Journal of Pathology. 2011;178(3):1092–105.

[59] Martiáñez T, Lamarca A, Casals N, Gella A. N-cadherin expression is regulated by UTP in schwannoma cells. Purinergic Signalling. 2013;9(2):259–70.

[60] Aslan E, Kocaeli H, Bekar A, Tolunay S, Ulus IH. CDP-choline and its endogenous metabolites, cytidine and choline, promote the nerve regeneration and improve the functional recovery of injured rat sciatic nerves. Neurological Research. 2011;33(7):766–73.

[61] Caner B, Kafa MI, Bekar A, Kurt MA, Karli N, Cansev M, Ulus IH. Intraperitoneal administration of CDP-choline or a combination of cytidine plus choline improves nerve regeneration and functional recovery in a rat model of sciatic nerve injury. Neurological Research. 2012;34(3):238–45.

[62] Pillois X, Chaulet H, Belloc I, Dupuch F, Desgranges C, Gadeau AP. Nucleotide receptors involved in UTP-induced rat arterial smooth muscle cell migration. Circulation Research. 2002;90(6):678–81.

[63] Ando RD, Mehesz B, Gyires K, Illes P, Sperlagh B. A comparative analysis of the activity of ligands acting at P2X and P2Y receptor subtypes in models of neuropathic, acute and inflammatory pain. British Journal of Pharmacology. 2010;159(5):1106–17.

[64] Parisi V, Coppola G, Ziccardi L, Gallinaro G, Falsini B. Cytidine-5′-diphosphocholine (Citicoline): a pilot study in patients with non-arteritic ischaemic optic neuropathy. European Journal of Neurology. 2008;15(5):465–74.

[65] Nunes CP, Higashi R, Ribeiro MG, de Souza Fonseca A, Leite A, Krymchantowski AV, et al. Efficacy and tolerability of a combination of uridine, cytidine, and vitamin B12 in anemia. A double-blind, comparative study versus nucleotide monotherapy. Revista Brasileira de Medicina. 2008;65(1/2):15.

[66] Nunes CP, Scussel Jr AB, Goldberg H, Goldwasser G, Oliveira L, Rzetelna H, et al. Alcoholic polyneuropathy: clinical assessment of treatment outcomes following therapy with nucleotides and vitamin B12. Research in Neurology: An International Journal. 2013;2013:g1–16.

[67] Roy RP, Ghosh K, Ghosh M, Acharyya A, Bhattacharya A, Pal M, Chakraborty S, Sengupta N. Study of vitamin B12 deficiency and peripheral neuropathy in metformin-treated early type 2 diabetes mellitus. Indian Journal of Endocrinology and Metabolism. 2016;20(5):631.

[68] Solomon LR. Functional vitamin B12 deficiency in advanced malignancy: implications for the management of neuropathy and neuropathic pain. Supportive Care in Cancer. 2016a;24(8):3489–94.

[69] Wang D, Zhai JX, Liu DW. Serum folate, vitamin B12 levels and diabetic peripheral neuropathy in type 2 diabetes: a meta-analysis. Molecular and Cellular Endocrinology. 2017. 163(3), 362–364.

[70] Aubart FC, Sedel F, Vicart S, Lyon-Caen O, & Fontaine, B. Neurological disorders due to vitamin B12 deficiency triggered by nitrous oxide. Neurological Review, 2007;163(3):362–4.

[71] Sakly G, Hellara O, Trabelsi A, & Dogui M. Reversible peripheral neuropathy related to vitamin B12 deficiency. Neurophysiologie Clinique/Clinical Neurophysiology. 2005;35(5):149–53.

[72] McCaddon A, Regland B, Hudson P, Davies G. Functional vitamin B12 deficiency and Alzheimer disease. Neurology. 2002;58(9):1395–9.

[73] Solomon LR. Vitamin B12-responsive neuropathies: a case series. Nutritional Neuroscience. 2016b;19(4):162–8.

[74] Roy S, Sable P, Khaire A, Randhir K, Kale A, Joshi S. Effect of maternal micronutrients (folic acid and vitamin B12) and omega 3 fatty acids on indices of brain oxidative stress in the offspring. Brain and Development. 2014;36(3):219–27.

[75] Guest J, Bilgin A, Hokin B, Mori TA, Croft KD, Grant R. Novel relationships between B12, folate and markers of inflammation, oxidative stress and NAD (H) levels, systemically and in the CNS of a healthy human cohort. Nutritional Neuroscience. 2015;18(8): 355–64.

[76] Petramfar P, Hosseinzadeh F, Mohammadi SS. Pseudo-Foster Kennedy Syndrome as a Rare Presentation of Vitamin B12 Deficiency. Iranian Red Crescent Medical Journal. 2016;18(6):e24610.

[77] Weir DG, Scott JM. Brain function in the elderly: role of vitamin B12 and folate. British Medical Bulletin. 1999;55(3):669–82.

[78] Jain P, Gulati S, Toteja GS, Bakhshi S, Seth R, Pandey RM. Serum alpha tocopherol, vitamin B12, and folate levels in childhood acute lymphoblastic leukemia survivors with and without neuropathy. Journal of Child Neurology. 2015;30(6):786–8.

[79] Morris N, Lynch K, Greenberg SA. Severe motor neuropathy or neuronopathy due to nitrous oxide toxicity after correction of vitamin B12 deficiency. Muscle & Nerve. 2015;51(4):614–6.

[80] Flippo TS, Holder WD. Neurologic degeneration associated with nitrous oxide anesthesia in patients with vitamin B12 deficiency. Archives of Surgery. 1993;128(12):1391–5.

[81] Stabler SP. Vitamin B12 deficiency. New England Journal of Medicine. 2013;368(2):149–60.

[82] Butler, CC, Vidal-Alaball, J, Cannings-John, R, McCaddon, A, Hood, K, Papaioannou, A, ... & Goringe, A. Oral vitamin B12 versus intramuscular vitamin B12 for vitamin B12 deficiency: a systematic review of randomized controlled trials. Family Practice. 2006;23(3):279–85.

[83] Gazoni FM, Malezan WR, Santos FC. B complex vitamins for analgesic therapy. Revista Dor. 2016;17(1):52–6.

[84] Dissanayake S. Assessing the bioequivalence of analogues of endogenous substances ('endogenous drugs'): considerations to optimize study design. British Journal of Clinical Pharmacology. 2010;69(3):238–44.

[85] Marzo A. Pharmacokinetics of endogenous substances. Bollettino Chimico Farmaceutico. 1992;131(5):181–4.

[86] Gasser T, Moyer JD, Handschumacher RE. Novel single-pass exchange of circulating uridine in rat liver. Science. 1981;213(4509):777–8.

[87] Monks A, Cysyk RL. Uridine regulation by the isolated rat liver: perfusion with an artificial oxygen carrier. American Journal of Physiology—Regulatory, Integrative and Comparative Physiology. 1982;242(5):R465–70.

[88] Pizzorno G, Cao D, Leffert JJ, Russell RL, Zhang D, Handschumacher RE. Homeostatic control of uridine and the role of uridine phosphorylase: a biological and clinical update. Biochimica et Biophysica Acta (BBA)-Molecular Basis of Disease. 2002;1587(2):133–44.

[89] Cansev M. Uridine and cytidine in the brain: their transport and utilization. Brain Research Reviews. 2006;52(2):389–97.

[90] Pastor-Anglada M, Felipe A, Casado FJ. Transport and mode of action of nucleoside derivatives used in chemical and antiviral therapies. Trends in Pharmacological Sciences. 1998;19(10):424–30.

[91] Lu H, Chen C, Klaassen C. Tissue distribution of concentrative and equilibrative nucleoside transporters in male and female rats and mice. Drug Metabolism and Disposition. 2004;32(12):1455–61.

[92] Redzic ZB, Biringer J, Barnes K, Baldwin SA, Alâ€ Sarraf H, Nicola PA, et al. Polarized distribution of nucleoside transporters in rat brain endothelial and choroid plexus epithelial cells. Journal of Neurochemistry. 2005;94(5):1420–6.

[93] Wurtman RJ, Regan M, Ulus I, Yu L. Effect of oral CDP-choline on plasma choline and uridine levels in humans. Biochemical Pharmacology. 2000;60(7):989–92.

[94] Lecca D, and Stefania C. Uracil nucleotides: from metabolic intermediates to neuroprotection and neuroinflammation. Biochemical Pharmacology. 2008;7510:1869–81.

[95] Wang, L, Pooler, AM, Albrecht, MA, & Wurtman, RJ. Dietary uridine-5′-monophosphate supplementation increases potassium-evoked dopamine release and promotes neurite outgrowth in aged rats. Journal of Molecular Neuroscience. (2005);27(1):137–45.

[96] Cansev M, and Wurtman RJ. Exogenous cytidine-5′-diphosphocholine increases brain cytidine-5′-diphosphocholine levels in gerbils. Journal of Neurochemistry. 2005;94:105–6.

[97] Savci, V, and Wurtman RJ. Effect of cytidine on membrane phospholipid synthesis in rat striatal slices. Journal of Neurochemistry. 1995;641:378–84.

[98] Kennedy EP, Weiss SB. The function of cytidine coenzymes in the biosynthesis of phospholipides. Journal of Biological Chemistry. 1956;222(1):193–214.

[99] Von Kügelgen I. Pharmacological profiles of cloned mammalian P2Y-receptor subtypes. Pharmacology & Therapeutics. 2006;110(3):415–32.

[100] Bevilacqua J, Downes C, Lowenstein P. Visualization of agonist-stimulated inositol phospholipid turnover in individual neurons of the rat cerebral cortex and hippocampus. Neuroscience. 1994;60(4):945–58.

[101] Park SY, Im Kim H, Shin YK, Lee CS, Park M, Song J-H. Modulation of sodium currents in rat sensory neurons by nucleotides. Brain Research. 2004;1006(2):168–76.

[102] Beamer E, Gölöncsér F, Horváth G, Bekő K, Otrokocsi L, Koványi B, Sperlágh B. Purinergic mechanisms in neuroinflammation: an update from molecules to behavior. Neuropharmacology. 2016;104:94–104.

[103] Jacobson, KA, and Müller CE. Medicinal chemistry of adenosine, P2Y and P2X receptors. Neuropharmacology. 2016;104:31–49.

[104] Boccazzi M, Rolando C, Abbracchio MP, Buffo A, Ceruti S. Purines regulate adult brain subventricular zone cell functions: contribution of reactive astrocytes. Glia. 2014;62(3):428–39.

[105] Burnstock G. An introduction to the roles of purinergic signalling in neurodegeneration, neuroprotection and neuroregeneration. Neuropharmacology. 2016;104:4–17.

[106] Fumagalli M, Lecca D, Abbracchio MP. CNS remyelination as a novel reparative approach to neurodegenerative diseases: the roles of purinergic signaling and the P2Y-like receptor GPR17. Neuropharmacology. 2016;104:82–93.

[107] Vrbova G, Mehra N, Shanmuganathan H, Tyreman N, Schachner M, Gordon T. Chemical communication between regenerating motor axons and Schwann cells in the growth pathway. European Journal of Neuroscience. 2009;30(3):366–75.

[108] Abbracchio MP, Burnstock G, Verkhratsky A, Zimmermann H. Purinergic signalling in the nervous system: an overview. Trends in Neurosciences. 2009;32(1):19–29.

[109] Förster D, Reiser G. Supportive or detrimental roles of P2Y receptors in brain pathology? – The two faces of P2Y receptors in stroke and neurodegeneration detected in neural cell and in animal model studies. Purinergic Signalling. 2015;11(4):441–54.

[110] Giaume C, Kirchhoff F, Matute C, Reichenbach A, Verkhratsky A. Glia: the fulcrum of brain diseases. Cell Death & Differentiation. 2007;14(7):1324–35.

[111] Guzman SJ, Gerevich Z. P2Y receptors in synaptic transmission and plasticity: therapeutic potential in cognitive dysfunction. Neural Plasticity. 2016;2016.

[112] Weisman GA, Camden JM, Peterson TS, Ajit D, Woods LT, Erb L. P2 receptors for extracellular nucleotides in the central nervous system: role of P2X7 and P2Y2 receptor interactions in neuroinflammation. Molecular Neurobiology. 2012;46(1):96–113.

[113] Weisman GA, Woods LT, Erb L, & Seye CI. P2Y receptors in the mammalian nervous system: pharmacology, ligands and therapeutic potential. CNS & Neurological Disorders-Drug Targets (Formerly Current Drug Targets-CNS & Neurological Disorders). 2012b;11(6):722–38.

[114] Kobayashi K, Yamanaka H, Yanamoto F, Okubo M, & Noguchi, K. Multiple P2Y subtypes in spinal microglia are involved in neuropathic pain after peripheral nerve injury. Glia. 2012;60(10):1529–39.

[115] Moriyama T, Iida T, Kobayashi K, Higashi T, Fukuoka T, Tsumura H, et al. Possible involvement of P2Y2 metabotropic receptors in ATP-induced transient receptor potential vanilloid receptor 1-mediated thermal hypersensitivity. Journal of Neuroscience. 2003;23(14):6058–62.

[116] Lecona E, Turnay J, Nieves O, GuzmÃ¡n-ArÃ¡nguez A, Morgan RO, Fernandez M-P, et al. Structural and functional characterization of recombinant mouse annexin A11: influence of calcium binding. Biochemical Journal. 2003;373(2):437–49.

[117] Abbracchio MP, Boeynaems JM, Barnard EA, Boyer JL, Kennedy C, Miras-Portugal MT, ... & Burnstock G. Characterization of the UDP-glucose receptor (re-named here the P2Y 14 receptor) adds diversity to the P2Y receptor family. Trends in Pharmacological Sciences. 2003;24(2):52–5.

[118] Chambers JK, Macdonald LE, Sarau HM, Ames RS, Freeman K, Foley JJ, ... & Trill J. AG protein-coupled receptor for UDP-glucose. Journal of Biological Chemistry. 2000;275(15):10767–71.

[119] Sponne IE, Gaire D, Stabler SP, Droesch S, Barbé FM, Allen RH, ... & Nicolas JP. Inhibition of vitamin B12 metabolism by OH-cobalamin c-lactam in rat oligodendrocytes in culture: a model for studying neuropathy due to vitamin B12 deficiency. Neuroscience Letters. 2000;288(3):191–4.

[120] Banerjee R, Ragsdale SW. The many faces of Vitamin B12: catalysis by cobalamin-dependent enzymes 1. Annual Review of Biochemistry. 2003;72(1):209–47.

[121] Clarke R, Smith AD, Jobst KA, Refsum H, Sutton L, Ueland PM. Folate, vitamin B12, and serum total homocysteine levels in confirmed Alzheimer disease. Archives of Neurology. 1998;55(11):1449–55.

[122] Justo R, Cesar M, Migowski E, Cisne R. Relation between vitamins of the b complex, GABA and glutamate, and their role in neurocognitive disorders-Brief review. International Journal of Basic and Applied Sciences. 2016;5(4):229–37.

[123] Ross J, Baldessarini MD. Neuropharmacology of S-adenosyl-L-methionine. The American Journal of Medicine. 1987;83(5):95–103.

[124] Surtees R. Biochemical pathogenesis of subacute combined degeneration of the spinal cord and brain. Journal of Inherited Metabolic Disease. 1993;16(4):762–70.

[125] Kumar N. Neurologic aspects of cobalamin (B12) deficiency. Handbook of Clinical Neurology. 2013;120:915–26.

[126] Fortun J, Verrier JD, Go JC, Madorsky I, Dunn WA, Notterpek L. The formation of peripheral myelin protein 22 aggregates is hindered by the enhancement of autophagy and expression of cytoplasmic chaperones. Neurobiology of Disease. 2007;25(2):252–65.

[127] Goldberg H, Júnior ABS, Cohen JC, Rzetelna H, Mezitis SGE, Nunes FP, Ozeri D, Daher JPL, Nunes CP, Oliveira L, Geller M. Neural compression-induced neuralgias: clinical evaluation of the effect of nucleotides associated with vitamin B12. Revista Brasileira de Medicina. 2009;66(11):380–5.

Chapter 6

The Role of Pharmacological Agents in Nerve Regeneration after Peripheral Nerve Repair

Agon Mekaj and Ymer Mekaj

Additional information is available at the end of the chapter

http://dx.doi.org/10.5772/intechopen.68378

Abstract

Peripheral nerve injuries are frequent and represent a significant pathology of the peripheral nervous system because, despite operative techniques and successful microsurgical repair, in most cases, the nerve repair is followed by scar formation. Numerous investigations have been carried out with the aim of finding pharmacological substances that can prevent scar formation and speed up the regeneration of repaired nerves. This chapter is dedicated to the efforts of many researchers to find different pharmacological agents with local effects on the improvement of nerve regeneration. Numerous experiments have been carried out in mice and rabbits using hyaluronic acid, tacrolimus, cyclosporin A, melatonin, vitamin B12, methylprednisolone, riluzole and potassium and calcium channel blockers. In the experimental animal studies, topical pharmacological agents were used at the site of peripheral nerve repair. The effect of these substances is most commonly studied in sciatic nerve injury in experimental animals. Their effects were evaluated using a variety of methods, such as morphological, biomechanical, electrophysiological and functional evaluation, and the above-mentioned substances, have been shown to have neuroprotective and neuroregenerative properties though different mechanisms.

Keywords: nerve injury, nerve regeneration, pharmacological agents, scar formation

1. Introduction

The peripheral nervous system (PNS) is very complex, being composed of the cranial nerves and the spinal nerves, which project from the spinal cord and pass through the intervertebral foramina of the vertebrae [1]. Peripheral nerves, composed of motor and sensory neurons, are considered as complex organs and are present in nearly all parts of the human body [2]. Motor neurons transmit processed information from the central nervous systems (CNS) to skeletal muscles via efferent pathways, whereas collected information from periphery travels to the brain via afferent pathways, after they are translated into nerve signals [3]. Neurons are the main cells of the PNS, but there are two kinds of neuroglia in the PNS, namely Schwann cells and satellite cells. Neurons are made up of the body (soma) and their thin processes of the cell, which are called dendrites and axons. In the soma of the neuron, there are many types of organelles [4]. Based on the number of extensions that arise from the cell body, neurons can be multipolar (three or more extensions), bipolar (two extensions, one axon and one dendrite) or unipolar (only one extension), which are very short and divided in the form of a letter T [5].

Peripheral nerve injuries (PNIs) are very common, and automobile accidents are the most common cause of nerve trauma [6], with most cases (75%) occurring in the upper limbs [7]. However, the etiological factors of PNI are different in peace and conflict periods, and, historically, most knowledge of PNIs was developed during wars [8]. Nerve injuries can be caused by lacerations with sharp objects, penetrating trauma, stretching or crushing trauma, fractures and wounds [9]. Lacerations, especially those which were caused by a knife blade, are another common cause of PNIs, comprising 30% of serious injuries in some series [10]. Compression is another common cause of PNIs, including 'Saturday Night Palsy' caused by radial nerve compression, which causes entrapment neuropathies [11]. The most severe form of nerve injury is a transection, which is known as grade V neurotmesis, usually owing to a laceration from a knife, firearm or glass shard [12]. The neurotmesis is characterised with a full transection of the axons and connective tissue layers wherein complete discontinuity of the nerve is observed [13]. There are numerous classifications of nerve injuries, but of these classifications, the most widely accepted are those developed by Seddon and Sunderland [14, 15].

After a peripheral nerve sustains a traumatic injury, complex pathophysiologic changes, such as morphologic and metabolic changes, occur at the injury site [16]. Furthermore, these complex changes occur in the nerve cell body, in the proximal and distal segments to the injury site, as well as in the distal endings of both muscle end-plates and sensory receptors [17]. These changes are characterised by axonal degeneration, which follows a sequence of events within the zone of trauma extending both proximally and distally [18]. Disconnected axons and cell bodies (in proximal axon injuries) degenerate via chromatolysis [19]. The degenerative changes in the distal segment were first described by Waller in 1850 based on observations of frog glossopharyngeal and hypoglossal nerves after injury [20]. Wallerian degeneration starts almost immediately after axotomy and lasts for 3–6 weeks [17].

The regenerative process begins almost immediately after nerve injury. The first wave of axonal sprouting occurs within hours of axotomy [21]. Two days after this first wave of axonal sprouting, a second wave of this process of regeneration starts [22]. According to some authors, axons may branch once they reach the distal stump, and in these cases, one axon may give rise to several branches [23]. It is known that Schwann cells play an important role in nerve regeneration at the site of nerve injury because they elaborate processes that include physical conduits that lead axons to their targets [17]. The extension of Schwann cells' processes can limit the rate of axon regeneration more than axonal growth [24]. Regeneration of the damaged peripheral nerve depends on the microsurgical procedure performed. Currently, there are several operating techniques that can be used to repair injured nerves, such as direct epineural repair, grouped fascicular repair, fascicular repair and nerve grafting [25]. However, there are some factors that influence the regeneration process after nerve repair, such as the nature, location and extent of damage, the extent and timing of repair, fascicular anatomy and patient factors (age, physical condition, metabolic disorders, avitaminosis and the presence of any disease).

In addition, recently some experimental studies have shown that nerve regeneration after its repair can be improved by some pharmaceutical agents, mainly used locally at the site of nerve repair. Drugs commonly used for this purpose include tacrolimus [26–29], hyaluronic acid and its derivatives [30–32], melatonin [33–35], methylprednisolone [36–39], vitamin B complex and vitamin B12, [40–42], calcium and potassium channel blockers [43, 44] and riluzole [45, 46]. These substances have neuroprotective and neuroregenerative properties, though different mechanisms contribute markedly to nerve regeneration.

Therefore, the main aim of this chapter is to present new insights into the mechanisms of action of many of the above-mentioned pharmacological agents on the prevention of perineural scar formation and on nerve regeneration after peripheral nerve surgery. However, it is understandable that complete regeneration and functional recovery will almost never be achieved, regardless of the operative technique used or the type of pharmacological agent applied.

2. Hyaluronic acid

Hyaluronic acid (HA) (CAS No. 9004-61-9) is a natural glycosaminoglycan formed by bonding *N*-acetyl-D-glucosamine with glucuronic acid [47]. It is a mucopolysaccharide, which occurs naturally in all living organisms, and is several thousands of sugars (carbohydrates) long. Disaccharide units are formed at the plasma membrane in vertebrates and some bacteria [48, 49]. HA is characterised by a very large number of disaccharide pairs (10,000 or more), so its molecular mass is approximately 4 million Da [50]. HA was discovered in bovine vitreous humour by Meyer and Palmer in 1934. These authors found that the HA contained two sugar molecules, one of which was uronic acid, and they proposed the name 'hyaluronic acid' [51], while the term 'hyaluronan' was introduced in 1986 by Endre Balazs to conform with the international nomenclature of polysaccharides [52]. HA is a major primary component in the extracellular matrix, but it has also been found intracellularly. HA has been isolated from many other sources, and its physicochemical structural properties and biological role have been studied in numerous laboratories [53]. The biosynthesis of HA has been studied for over six decades, but our understanding of the biochemical details of HA assembly is still incomplete. The enzyme responsible, HA synthase (HAS), is a membrane protein that requires only Mg^{2+} and two sugar-UDP substrates (GlcUA-UDP and GlcNAc-UDP) to polymerise HA chains [54]. In 1993, the hasA gene was identified and cloned, and the HAS protein from *Streptococcus pyogenes* was expressed [55, 56]. It was also demonstrated that only the HAS protein was required to synthesise HA [57]. It is known that mammalian genomes have three different HAS genes (HAS1, HAS2 and HAS3) that are expressed at specific times and in specific tissues during development, ageing and wound healing, as well as under normal and some pathologic conditions [58, 59].

HA has numerous biological functions, such as maintenance of the elastoviscosity of liquid connective tissues, for example, in joint synovial and eye vitreous fluid, control of tissue hydration and water transport, as well as supramolecular assembly of proteoglycans in the extracellular matrix. Furthermore, HA has various receptor-mediated roles in cell detachment, mitosis, migration, tumour development and metastasis, and inflammation [52, 60]. The predominant role of HA in organisms is unknown, but some clinical studies have demonstrated various physiological effects of exogenous HA. Exogenous HA enhances chondrocyte HA and proteoglycan synthesis, reduces the reproduction and activity of proinflammatory mediators, such as matrix metalloproteinases, and alters the behaviour of immune cells [61]. HA has also been successfully used in peripheral nerve surgery to reduce nerve adhesions during wound healing after nerve injury, which occur during ophthalmological, cardiovascular and dermatological procedures, including supplementing joint fluid in arthritis [32, 62, 63]. HA is known to reduce the extent of scar formation and nerve adhesions via the inhibition of lymphocyte migration, proliferation and chemotaxis of granulocyte phagocytosis and degranulation, and macrophage motility in order improve peripheral nerve regeneration [31, 64]. These functions are manifested during scavenging of reactive oxygen-derived free radicals, the inhibition of immune complex adherence to polymorphonuclear cells, the inhibition of leucocyte and macrophage migration and aggregation, and the regulation of fibroblast proliferation [65]. HA is an endogenous stimulator of interleukin-1 (IL-1) production, and IL-1 affects fibroblasts proliferation and collagenase production [30]. Therefore, according to Hiro et al., HA is an endogenous IL-1 inducer and may play important roles in the pathological and/or physiological changes of connective tissues [30]. It is known that HA is highly non-antigenic and non-immunogenic, because it has high structural homology across species and weak interactions with blood components [66]. HA's degradation products are thought to contribute in scar formation because the increased amounts of HA fragments from the action of hyaluronidase in HA induce increased scar formation. There are various commercial preparations of HA in different forms, such as films, microspheres, liposomes, fibres and hydrogels, which have been used for more than 20 years worldwide [63]. Although the above-mentioned commercial preparations of HA have mainly been used in animal studies, it also provides useful information regarding the effect of hyaluronate in the prevention of post-operative peridural scar adhesion after laminectomy in spine surgery; however, additional clinical trials regarding the use of HA-based gels should be performed to confirm its effects in human subjects [67, 68]. Use of the hyaluronic acid-carboxymethylcellulose membrane Seprafilm as a solid anti-adhesion barrier agent is one of the therapeutic approaches used to reduce postoperative scar formation and is effective in promoting peripheral nerve regeneration at the repair site [69]. In addition, HA-carboxymethylcellulose solutions improve nerve regeneration and reduce perineural scar formation and adhesion after sciatic nerve repair [70]. It has been confirmed that direct application of HA-carboxymethylcellulose in transected nerves may limit axonal outgrowth by contact with regenerating axons; therefore, HA-carboxymethylcellulose barriers may prove to be a tool to prevent neuroma formation through inhibiting axonal growth [71]. On the contrary, according to some other studies, the role of HA solution in axonal outgrowth is dose-dependent, because high dose of HA (100–1000 μg/ml) topically used is characterised by significantly increased axonal outgrowth compared with HA solution (10 μg/ml) applied in the control group, in which axonal outgrowth did not occur [72]. Some authors describe the early effect on nerve regeneration of continuous local delivery of nerve growth factor (NGF) and the local incorporation of HA inside a newly manufactured nerve conduit material from fresh human amniotic membrane [73]. Additionally, other authors have demonstrated that the combination of vascular endothelial growth factor (VEGF) gene therapy and a HA film sheath-enriched microenvironment may synergistically promote peripheral nerve regeneration [74]. The microenvironment of neuron cells plays a crucial role in regulating neural development and regeneration [75].

HA and its derivatives may also promote regeneration of injured nerves through realignment of the fibrin matrix, and they can provide a suitable environment for axonal ingrowths [25]. In another animal experimental study, a single topical dose of HA enhanced the nerve regeneration process in hindlimb rat and rabbit models by preventing perineural scar formation after peripheral nerve repair [31, 76]. New data in the literature have shown that HA-based biomaterials have been applied in a wide range of medical and biological fields and play important roles in neural regeneration [75].

3. Tacrolimus

Tacrolimus, also known as FK506 or Fujimicin (C44H69O12), is a macrolide immunosuppressive drug that is approved for the prevention of allograft rejection [25], but it is well known that tacrolimus can also increase nerve regeneration and facilitate allografting of nerves via immunosuppression [18]. Chemically, tacrolimus is a 23-membered macrolide lactone. It is a powerful and selective anti-T-lymphocyte agent that was discovered in 1984 and later approved by the U.S. Food and Drug Administration (FDA). This agent, isolated from the fungus *Streptomyces tsukubaensis*, has a mechanism of action similar to that of cyclosporin A. The first preliminary report on FK506 was presented in 1986 at the 11th International Congress of the Transplantation Society [77], and the first experimental reports were published in 1987 [78–80]. These early reports demonstrated that FK506 is a potent immunosuppressant that acts *in vitro* via inhibiting interleukin 2 (IL-2) production, as well as by inhibiting the response of mixed lymphocyte cultures at concentrations 32–100 times lower than that of cyclosporin A [81]. It means that despite similar mechanisms of action, tacrolimus is 50–100 times more potent than cyclosporin A [82].

Tacrolimus is able to modulate the immune system, inhibit T-cell function by binding to FK binding proteins (FKBP) and mediate immunosuppression by inhibiting calcineurin and calcium and calmodulin-dependent phosphatase. The primary biological effect of calcineurin inhibition includes the decrease of the production of inflammatory cytokines such as tumour necrosis factor (TNF)-α, interleukin-2 and interferon-γ [28]. The drug's immunosuppressive effects are mediated largely through FKBP12, which is involved in intracellular calcium flux and cycle regulation [83]. Tacrolimus realises its effect by binding to its receptors (FKBP12 and FKBP52). The FKBP12 receptors are responsible for immunosuppressive effects, whereas the FKBP52 receptors are related to neuroregenerative effects. These effects of tacrolimus have been shown experimentally in multiple models of nerve injury during the past decade when tacrolimus was used in sub-immunosuppressive doses, and these findings have stimulated interest in characterising its neurophysiologic effects on nerve regeneration [84]. Tacrolimus sustains nerve regeneration with both systemic and local administration [85]. *In vitro* and *in vivo* experimental models have proven that tacrolimus increases neurite elongation and accelerates the rate of peripheral nerve regeneration [29]. In addition, there is evidence from *in vivo* experimental studies that highlight that tacrolimus has a neuroprotective role in the central nervous system through its direct impact on its various cell populations [86]. Recently, some studies have demonstrated the effect of tacrolimus in spinal surgery where it was found that topical application of tacrolimus could inhibit fibroblast proliferation and prevent epidural scar adhesion after laminectomy in a rat model [87], and it is worth mentioning that tacrolimus has been successfully used topically in spinal cord trauma for neuroprotection and local regeneration [88].

Tacrolimus is an essential drug for the conventional immunosuppression regimen for solid organ transplantation. The use of tacrolimus in post-transplant immunosuppressive regimens can enhance nerve regeneration and the growth of axon sprouts into donor tissue [89]. It has been demonstrated that tacrolimus has a powerful effect on promoting axon regeneration through its immunosuppressive and neurotrophic action [90]. This (neurotrophic) action can be completely prevented *in vitro* by the addition of a monoclonal antibody against FKBP52 [84].

The topical effects of tacrolimus on peripheral nerve have not been well investigated to date and the exact mechanism by which tacrolimus affects nerve regeneration is unclear, but outcomes data, so far, have been promising [89]. Additionally, the results of the use of tacrolimus in peripheral nerve regeneration differ in the literature. The relative variability of the results of experimental studies of nerve injures can be explain by the variety of models and testing methods used [91]. Prior studies have shown that FK506-FKBP12 interaction may lead to a neuroregenerative effect through increased neuronal expression of a growth cone-associated protein GAP-43, but there is evidence that this occurs through inactivation of neuronal nitric synthetase [92, 93]. Moreover, axon regeneration from tacrolimus is realised predominantly through its binding to FKBP-12, which activates GAP-43 and the transforming growth factor (TGFb1) pathway [26]. Tacrolimus can promote peripheral nerve regeneration through reducing scar formation; however, little is known about how tacrolimus reduces scar formation [94, 95]. Que et al. suggest that tacrolimus-induced fibroblast apoptosis contributes to the suppression of fibroblast proliferation and then causes the reduction of scar formation in the damaged nerve; in fibroblasts, apoptosis of tacrolimus involves c-Jun N-terminal kinase (JNK) and extracellular-signal-regulated kinase [94].

Some studies have shown a positive effect of tacrolimus after it was used in different allograft and isographs. Earlier axon regeneration in allografts with FK506 compared to allografts without FK506 was demonstrated experimentally [96]. Systemic application of tacrolimus at doses of 0.6 mg/kg found the amount of myelin debris in autologous nerve grafts to be decreasing [97]. This can be explained by the reduction of macrophage infiltration after tacrolimus administration. The reduction of scar formation at the site of nerve repair by the above-mentioned mechanisms has been associated with better morphologic and nerve function recovery. Recently, we published two original articles that compare the effects of HA and FK-506 on nerve regeneration. We found that our electrophysiological and biomechanical measurements as well as our results of functional evaluations indirectly indicate that the effects of HA and FK506 on nerve regeneration are similar [98, 99].

4. Cyclosporin A

Cyclosporin A is a neutral lipophilic cyclic undecapeptide that was isolated in 1971 from the fungus *Tolypocladium inflatum* and came into medical use in 1983 [100]. Its immunosuppressive function was first reported in 1972 by Sandoz Laboratories [101]. Discovery of cyclosporine A has revolutionised transplantation medicine, which requires the application of immunosuppressive therapy. Cyclosporin has been widely applied as an immunosuppressive substance in organ transplantation in association with other drugs, both in vital and non-vital organs, such as skin, nerve and muscles. It is known that cyclosporin A has similar immune-suppressing characteristics to tacrolimus, but tacrolimus has a more potent effect with equal volumes of drug [95]. There are data in the literature that cyclosporin A was used experimentally in order to investigate its anti-scarring effects on peripheral nerves both ultrastructurally and in gross post-surgical and histopathological analyses [102]. Cyclosporin's mechanism of action in nerve regeneration remains controversial [103]. However, probable mechanisms include inhibiting white blood cell proliferation and/or differentiation and inhibiting Ca^{2+}-dependent cell injury [104]. Besides the above-mentioned mechanism, cyclosporin A acts with other anti-inflammatory effects in preventing scar formation because it can block the transcription of cytokine genes in activated T cells, whereas on the another hand, it is well established that cyclosporin A inhibits the phosphatase activity of calcineurin through the formation of a complex with cyclophilin, which regulates nuclear translocation and subsequent activation of nuclear factor of activated T cells (NFAT) transcription factors [105]. Calcineurin plays an important role in the T-cell receptor-mediated signal transduction pathway and is identified as the common target for cyclosporin A and tacrolimus [106].

The role of cyclosporin A in peripheral nerve regeneration after peripheral nerve allografting has been investigated in experimental models immunosuppressed with cyclosporine for more than two decades [107]. It is worth mentioning that most of these studies were concentrated on allograft survival, rather than on the direct effect of cyclosporin A on peripheral nerve regeneration [103]. Some authors have investigated the efficacy of cyclosporine A in large- and small-diameter nerve grafts as well as in long and short allografts. They have found better nerve regeneration in large-diameter nerve grafts than in small-diameter nerve grafts, whereas with regard to the length of the grafted nerve, short nerve allografts give higher axon counts than long ones, the same as with autografts [108]. Recently published data suggest that even though cyclosporin A is effective at reducing graft rejection, axon regeneration is still superior in autografts versus immunosuppressed allografts [109]. Furthermore, cyclosporin A effectively prevented postoperative epineurial fibrosis on rat sciatic nerves after peripheral nerve surgery with no adverse effects after topical application [102]. The application of cyclosporin A in a silicon conduit neurorrhaphy resulted in improvement of functional recovery and quantitative morphometric indices of sciatic nerves in diabetic rats [110].

5. Melatonin

Melatonin, which is also known as *N*-acetyl-5-methoxytryptamine, is a hormone and was first identified in bovine pineal extracts [35]. Melatonin is the main hormone of the pineal gland and is an important signalling molecule that occurs in many organisms as well as in plants and fungi [33]. The pineal gland is in the middle of the brain and secretes melatonin, a hormone that regulates when you sleep at night and wake up in the morning, as well as other numerous aspects of circadian biology [111]. Melatonin has an effect on the morphologic features of the nerve tissue, suggesting its neuroprotective, free-radical scavenging and antioxidative and analgesic effects in degenerative diseases of peripheral nerves [25]. There are different opinions among authors regarding the protective effect of melatonin in stimulation of peripheral regeneration, because some authors have reported toxic effects of melatonin on peripheral nerves [33]. However, nowadays there is enough evidence from the literature showing that melatonin has a useful effect on axon length and sprouting after traumatic events to peripheral nerves [34]. The beneficial effects of melatonin administration on the recovery of injured nerves may be attributed to its antioxidant properties [112]. Melatonin has an effect on superoxide dismutase, which is an important antioxidative enzyme that is involved in redox regulation of regulative stress, and would exert melatonin's beneficial effects by preserving the superoxide dismutase reactivity following peripheral nerve injury [113]. The rhythm of melatonin defines the activity of glutathione peroxidase and consequently also glutathione reductase. It is thought that the involvement of melatonin in the control of redox processes depends on its high-affinity binding to cytosolic quinone reductase 2, previously believed to be a melatonin receptor [114]. Through a variety of experimental neuropathologies involving nitric oxide (NO), it was confirmed that melatonin exerts its neuroprotective role after peripheral axotomy via reduction of oxidative damage [115]. The neuronal isoform of nitric oxide synthetase (nNOS), an NADPH-dependent diaphorase, is considered to play a role in motoneuron death induced by nerve transection. In addition, it is known that exogenous melatonin can prevent neuropathy development via the inhibition of lipid peroxidation in renal tissue and the inhibition of TGF-β, which limits the effects against fibrosis [116]. Furthermore, data show that melatonin can significantly promote Schwann cell proliferation and can improve nerve regeneration after peripheral nerve injury via this mechanism both *in vitro* and *in vivo*. The functional recovery of damaged nerves was estimated by the amount of Schwann cells and the number of re-innervated muscle motor end-plate targets [117]. Furthermore, it is worth mentioning that Turgut et al. have experimentally demonstrated in rats that melatonin prevents neuroma formation after transacting the sciatic nerve by enhancing axonal regeneration [118]. Recently, some studies have shown the positive effect of melatonin on preventing scar formation, increasing nerve regeneration and improving functional recovery [119, 120]. However, the exact mechanisms by which melatonin limits fibrosis are currently unclear.

6. Methylprednisolone

Methylprednisolone is an anti-inflammatory pharmacological agent that has found widespread use in treatment of many pathological disorders in humans. It has also been experimentally investigated intensely for preventing scar formation and nerve regeneration because it is considered to have a neuroprotective role. Generally, glucocorticoids are anti-inflammatory substances that are often used to alleviate tissue oedema and trauma-induced inflammatory response because they can down regulate the expression of pro-inflammatory factors, such as tumour necrosis factor-α and interleukin-1β [121]. These pro-inflammatory factors can increase the expression of induced nitric oxide synthase in the injured region, leading to nitric oxide production and cell apoptosis [122]. There are some mechanisms by which glucocorticoids can express their anti-inflammatory effects in central and peripheral nerve system. However, one possible mechanism by which methylprednisolone inhibits nerve inflammation is its inhibition of CD3-positive inflammatory cell infiltration of local tissue [123]. It was found that higher doses of methylprednisolone have a neuroprotective role in injured nerves through inhibition of oxygen-free radical-induced lipid peroxidation [124]. Therefore, through inhibition of lipid peroxidation, methylprednisolone can retard both anterograde and retrograde nerve degeneration after peripheral nerve injury. Moreover, the steroid expresses its anti-inflammatory effects via inhibition on responsive cells and consequently recruitment of macrophages [125]. Regarding these anti-inflammatory effects, inhibition of phospholipase A2 activity should be mentioned, along with prevention of granulocyte, mast cell and macrophage degranulation, inhibition of macrophage migration-inhibitory factor and stabilisation of the lysosomal membrane, which are beneficial for treating injured nerves [39]. In the same study, the effect of preoperative locally administered dexamethasone on the recovery of crushed nerves was examined, and the authors concluded that local dexamethasone is more effective than systemic dexamethasone [39]. Systemic application in rats of moderate doses of methylprednisolone, i.e. 15–30 mg/kg, can effectively increase peripheral nerve regeneration, and it was also found that local administration of the drug can have the same positive effects as those of systemic administration while reducing systemic side effects [126]. Moreover, it was found that dexamethasone loaded in silicone tubes can improve functional recovery and morphometric indices of the sciatic nerve, and it was confirmed that topical administration of dexamethasone on peripheral nerve offers the benefits of cost savings as well as avoiding the complications associated with systemic administration [38]. The effect of methylprednisolone in suppressing scar formation and improving axonal regeneration after transection and suture of rat peripheral nerves was described many years before in rats [36].

Recent experimental studies have demonstrated that topical application of methylprednisolone can be realised using various methods, for example, in various materials such as silicon tubes [38], amniotic membranes [125] and microsphere sustained-release membranes [126], in order to avoid the rapid destruction of methylprednisolone at the site of nerve repair.

7. Vitamin B12

Vitamin B12, also called cobalamin, is a water-soluble vitamin with multiple functions in organisms, although in comparison with other nutrients, the body needs them in relatively small amounts. It is naturally present in animal products, fish, meat, poultry, eggs, milk and milk products [127]. There are several forms of vitamin B12, which contains mineral cobalt, so for this reason, compounds with vitamin B12 activity are collectively called cobalamins and the active forms are methylcobalamin and 5 deoxyadenosylcobalaminin [128]. Vitamin B12 as a coenzyme induces conversion of homocysteine to methionine in order to facilitate synthesis of nucleic acids and proteins. Therefore, it accomplishes the following essential nerve function, such as promotion of nerve regeneration owing to axoplasm flow within the neuraxon, in order to normalise the neuraxon's skeleton protein transportation as well as accelerate the formation of the myelin sheath [129]. Vitamin B12 in combination of B1 (thiamine) and B6 (pyridoxine) reduced degenerating processes in the nervous system, and therefore, this combination has been clinically administered [130]. Furthermore, this vitamin is involved in the metabolism of every cell in the human body, especially affecting DNA synthesis and fatty acid and amino acid metabolism [131]. It is known that B12 deficiency leads to deficiency in methionine, which is required for the synthesis of both phospholipids and myelin; therefore, it is an essential element in the maintenance of nerve functions because it induces synthesis of the myelin sheath and improves nerve conduction velocity. In addition, it was found that vitamin B12 increased the number of Schwann cells and myelinated nerve fibres, and the diameter of axons, through which effects it can promote the regeneration of myelinated nerve fibres and the proliferation of Schwann cells [132]. In addition, vitamin B12 has shown antioxidant properties because it is also a good scavenger of reactive oxygen species and is suggested to be a good neuroprotectant. Moreover, vitamin B complex or vitamin B12 can increase the expression of brain-derived neurotrophic factor (BDNF) in injured nerves at both mRNA and protein levels, therefore promoting the regeneration and functional recovery of injured nerves through increasing BDNF expression [41]. Some authors have shown that vitamin B12 provides a basis for more beneficial treatments of nervous disorders through both systemic and local delivery of high doses of methylcobalamin to target organs, which has been shown to have the potential to treat peripheral nerve injury [40]. Inasmuch as the amount of vitamin B complex and vitamin B12 vary in cases of crush nerve injuries, it is necessary to administrate these vitamins in the acute phase of nerve injury in order to enhance nerve regeneration [42].

8. Riluzole

Riluzole (2-amino-6-trifluoromethoxy-benzothiazol) is a benzothiazole anti-convulsant and the only U.S. Food and Drug Administration (FDA)-approved drug to treat amyotrophic lateral sclerosis (ALS) [133]. Riluzole is a sodium/glutamate antagonist that has been shown to have a neuroprotective effect, recently entered clinical testing for spinal cord injury [134] and currently is under Phase III clinical trial for the treatment of spinal cord injury (ClinicalTrials.gov: NCT01597518) [135]. Its neuroprotective effects are a result of the blockade of sodium channels and, consequently, prevention of Ca^{2+} overflow [136]. In experimental trials in animal models, it was successfully used to reduce symptoms in neurodegenerative disease and neural tissue injury, and these effects can be explained by its inhibition of presynaptic glutamate release through blocking voltage-gated sodium channels [133]. It is known that *in vitro* application of riluzole to adult dorsal root ganglion neurons gives a neuroprotective effect via promotion of neurite outgrowth in terms of number, length and branch [137]. For nerve regeneration after nerve injury, neurite outgrowth of surviving neurons is very important in order to reinervate target tissue. In addition, riluzole inhibits neuro-excitotoxicity in animal models of neural injury, and soon after its administration, it can sufficiently reduce pain from nerve root compression and can prevent development of neuronal dysfunction in the nerve root and the spinal cord [138]. Recently, riluzole was clinically approved for the treatment of motor neuron disease, and experimental research is now underway for the assessment of its role on nerve regeneration processes after peripheral nerve injury [46].

9. 4-Aminopyridine

4-Aminopyridine (4-AP) is a potassium channel blocker with the chemical formula $C_5H_4N{-}NH_2$ that it used as a research tool in order to classify the subtypes of the potassium channel [139]. It has shown clinical efficacy in the treatment of neurological disorders such as multiple sclerosis [140]. The mechanisms of action of 4-AP can explain by its effect in allowing impulse conduction in demyelinated axons by blocking K^+ channels that allow leakage of K^+ from these axons and thereby enabling axons to restore the level of depolarisation required for propagation of action potentials [141]. Recently, there are data from literature that 4-aminopyridine is a potent small molecule with neuroregenerative properties that enhances both the speed and extent of functional recovery after acute peripheral nerve injury, because it promotes remyelination [142]. The same authors have found that 4-aminopyridine treatment enables differentiation between incomplete and complete lesions more rapidly compared with existing approaches [142].

10. Verapamil

Verapamil belongs to the class of medications called calcium channel blockers. Besides its effects in the cardiovascular system, recently some experimental studies have investigated the role of verapamil in the peripheral nervous system, and it has been shown to reduce scar formation through inhibiting fibroblast adhesion and proliferation *in vitro* [143]. However, it is not clear whether topical application of verapamil after surgical nerve repair *in vivo* could prevent scar formation and promote nerve regeneration [44]. Apparently, this role of verapamil consists of stimulation of the endogenous anti-inflammatory reaction and decreasing pro-inflammatory processes by a channel blocker, therefore causing pain modulation or nerve regeneration [43]. The effect of calcium channel blockers in the reduction of scar formation was first reported by Lee and Ping [144]. There are two mechanisms by which verapamil can prevent scar formation: by reducing the biological activity of cells through inhibiting signal transduction inside and outside fibroblasts, and by suppressing the synthesis and secretion of collagen and extracellular matrix through changing fibroblast morphology [44].

The overall effects and mechanisms of the above-mentioned pharmacological agents in the prevention of scar formation and improved nerve regeneration are presented in **Table 1**.

Pharmacological agents	Effects	Mechanisms of action	References
Hyaluronic acid	Reduce the extent of scar formation and nerve adhesions	Via proliferation and chemotaxis of granulocyte phagocytosis and degranulation, and macrophage motility	Ozgenel [31] and Park et al. [70]
		Stimulator of interleukin-1 (IL-1) production, which affects (decreases) fibroblast proliferation and collagenase production	Hiro et al. [30]
Tacrolimus (FK506)	Neuroprotective role via reduction of scar formation	Through encouraging fibroblast apoptosis, it contributes to the suppression of fibroblast proliferation. Fibroblast apoptosis by tacrolimus involves c-Jun N-terminal kinase (JNK) and extracellular-signal-regulated kinase	Que et al. [94]
	Neuroregenerative role	FK506-FKBP12 interaction may lead to a neuroregenerative effect through increased neuronal expression of growth cone-associated protein GAP-43 that probably occurs through inactivation of neuronal nitric synthetase	Dawson et al. [92] and Madsen et al. [93]
		By increased neurite elongation and accelerating the rate of nerve regeneration	Konofoas and Terzis [29]
Cyclosporin A	Anti-scarring effects and nerve regeneration on peripheral nerves	Probable mechanisms include inhibiting white blood cell proliferation and/ or differentiation and inhibiting Ca^{2+}-dependent cell injury	Erkutlu et al. [104]
		It is well established that cyclosporin A inhibits the phosphatase activity of calcineurin through the formation of a complex with cyclophilin, which regulates nuclear translocation and subsequent activation of nuclear factor of activated T cells (NFAT) transcription factors	Matsuda and Koyasu [105]
Melatonin	Induces axon length and sprouting after traumatic events to peripheral nerves	Via its effect on superoxide dismutase, which is an important antioxidative enzyme that is involved in redox regulation of regulative stress	Chang et al. [113]
	Improves nerve regeneration	Via reduction of oxidative damage, melatonin exerts its neuroprotective role after peripheral axotomy in a variety of experimental neuropathologies that involve nitric oxide (NO)	Chang et al. [115]
	Limits fibrosis and neuroma formation	Through promoting Schwann cell proliferation	Chang et al. [117]
		Exact mechanisms through which melatonin imparts these effects are currently unclear	Turgut et al. [118]

Pharmacological agents	Effects	Mechanisms of action	References
Methylprednisolone	Neuroprotective role	By its inhibition of CD3-positive inflammatory cell infiltration in local tissue and consequently recruitment of macrophages	Feng and Yuan [123]
	Anti-inflammatory effects	Through inhibition of oxygen-free radical-induced lipid peroxidation	Hall [124]
Vitamin B12	Nerve regeneration	Owing to axoplasm flow within the neuraxon, in order to normalise the neuraxon's skeleton protein transportation as well as accelerating the formation of the myelin sheath	Wang et al. [129]
	Neuroprotectant	Through increasing the number of Schwann cells and myelinated nerve fibres, and the diameter of axons	Lopatina et al. [132]
		Through antioxidant properties, because it is also a good scavenger of reactive oxygen species	Sun et al. [41]
Riluzole	Neuroprotectant	Through blockade of sodium channels, and consequently prevention of Ca^{2+} overflow	Fehlings et al. [136]
	Inhibits neuro-excitotoxicity	Through reducing pain from nerve root compression, it can prevent development of neuronal dysfunction in the nerve root	Nicholson et al. [138]
4-Aminopyridine (4-AP)	Neuroregenerative properties	Through its effect in allowing impulse conduction in demyelinated axons by blocking K^+ channels that allow leakage of K^+ from these axons and thereby enabling axons to restore the level of depolarisation required for propagation of action potentials	Hayes [141]
Verapamil	Reduce scar formation and promote nerve regeneration	There are two mechanisms by which verapamil can prevent scar formation: by reducing the biological activity of cells through inhibiting signal transduction inside and outside fibroblasts, and by suppressing the synthesis and secretion of collagen and extracellular matrix through changing fibroblast morphology	Han et al. [44]

Table 1. The effects and mechanisms of pharmacological agents in nerve regeneration.

11. Discussion

There have been many efforts to diminish scar formation and perineural adhesion as well as to improve nerve regeneration after microsurgical nerve repair. Various surgical techniques and several different pharmacological agents have been used for this purpose. The evaluation of the effects of these pharmacological agents in the prevention of scar formation and nerve regeneration in experimental animals is performed by electrophysiological measurements and through assessing functional recovery, whereas after sacrificing the animals, other methods have been used, such as macroscopic, histomorphometric, immunohistochemical and biomechanical techniques [70, 91, 98, 99]. Ozgenel found that nerves treated with HA have a significant reduction in perineural thickness compared to nerves treated with just saline ($P < 0.05$). Besides that, this author found better mean conduction velocities (MCVs) and faster functional recovery in HA-treated nerves (0.82 ± 0.08 m/s) compared with nerves treated with saline, in which the MCV was 0.76 ± 0.04 m/s ($P < 0.05$) [31]. Park et al. found that topical application of HA carboxymethylcellulose solutions in rats significantly reduced nerve adherence score and the number of cellular components compared with the saline group (control group) ($P < 0.05$) [70]. The authors concluded that HA carboxymethylcellulose solutions improved nerve regeneration and reduced perineural scar formation and adhesion after sciatic nerve repair [70]. Furthermore, the same results were demonstrated by Adanali et al. in rabbit sciatic nerves where they used HA carboxymethylcellulose membranes in the experimental group and saline in the control group; they observed that adhesion in the surrounding tissues was significantly less in the HA carboxymethylcellulose membranes group than in the saline group [69]. Ikeda et al. found that local application of HA in the sciatic nerve was the most effective at reducing extraneural and intraneural connective tissue, compared with the steroid and saline groups [145]. By electrophysiological measurements, Ikeda et al. also found that the latencies of the HA and steroid groups were much shorter than that of the neurolysis group (2.14 ± 0.20, 1.92 ± 0.11 and 1.91 ± 0.15 m/s, respectively), but longer than that of the control group (1.68 ± 0.07 m/s). Similar results were found by histological examination, because scar tissue in the neurolysis group was thicker and more voluminous than that in the HA group or the steroid group [145]. In addition, Zor et al. found significantly less scar formation ($P < 0.01$) and significantly higher peak amplitudes in rats ($P < 0.01$) that received a combination treatment of vascular endothelial growth factor gene therapy with HA [74].

Shahraki et al. demonstrated earlier axon regeneration in allografts with FK506 compared to allografts without FK506 ($P < 0.05$) [96]. Yan et al. found that short treatment courses of 10 and 20 days with FK506 (in the graft model) were sufficient to reduce functional recovery time by 15 and 21%, respectively, compared with negative controls assessed by walking track analysis [90]. In addition, via a functional study, Azizi et al. confirmed faster recovery of the regenerated axons in the inside-out vein graft/FK506 group than that for the inside-out vein graft without FK506 (control group) ($P < 0.05$). The same statistically significantly difference was found when comparing these groups regarding the mean gastrocnemius muscle weight ratio ($P < 0.05$) [91]. Que et al. showed that scar area had a significant positive correlation with the fibroblast number, as detected by linear correlation analysis [94]. Other authors reported that tacrolimus can increase the number of axons and their myelinated axons by 40% and reduce by half the time to neurological recovery [29]. Furthermore, Li et al. found that after application of FK506 loaded in a chitosan guide, the amplitude and velocity of compound muscle action potential (CMAP) reached 60 and 73% of the control values, respectively [146].

In order to compare the effects of HA and FK506 on peripheral nerve regeneration in rabbits after the drugs were topically applied at the site of sciatic nerve, we used electrophysiological, macroscopic and microscopic methods, while functional assessment was performed via toe-spreading reflex. According to our results, HA and FK506 appear to have similar effects ($P > 0.05$) with respect to preventing scar formation and improving nerve regeneration compared with saline ($P < 0.05$) [80, 98]. However, it should be mentioned that we observed no significant differences in biomechanical properties in the HA and FK506 groups compared to the saline group ($P > 0.05$) [99].

Çetinalp et al. demonstrated that animals treated with cyclosporin A had statistically significant lower perineural adhesion and better separability than the saline group (control group) ($P < 0.0001$) [102]. However, these authors did not find a significant difference in the wound-healing characteristics or neurological functions between the treatment (cyclosporin A) group and the control group ($P > 0.05$) [102]. In an experimental rat sciatic nerve injection injury model established by penicillin G potassium injection, Erkutlu et al. randomly divided rats into three groups based on the length of time after nerve injury induced by cyclosporin A administration (30 minutes, 8 or 24 hours), recorded electrophysiological measurements (compound muscle action potentials, pre-injury, early post-injury [within 1 hour]and 4 weeks after injury) and then compared the results of the experimental groups with the control group. Finally, they found significant improvement of the compound muscle action potential amplitude value only when cyclosporin A was administered within 30 minutes of the injection injury ($P < 0.05$) [104].

Turgut et al. examined the gross morphology of neuroma formation in the proximal nerve segment via macroscopic and microscopic findings, and the surgical pinealectomy group without application of melatonin caused a proliferation of connective tissue and large neuroma formation at the proximal ends of transacted nerves compared with the surgical pinealectomy group and the group given melatonin ($P < 0.005$) [118]. In addition, Kaya et al. demonstrated a beneficial effect on axonal regeneration and functional recovery in the experimental group in which melatonin was applied after stripping of the epineurial vessels compared with other groups without melatonin application [119].

Recently, Sadraie et al. found that at 8 weeks after surgery, sciatic functional index, withdrawal reflex latency test, electrophysiological values and histological results in the amniotic membrane with the betamethasone group were improved compared to those in the control and sham groups ($P < 0.05$) [125]. Furthermore, Feng and Yuan found better and faster functional recovery in the dexamethasone-administered group compared to other groups without dexamethasone ($P < 0.05$); therefore, they concluded that dexamethasone can promote functional recovery after sciatic nerve crush injury [123]. In addition, Sun et al. via morphological (by electron microscopy) and functional analysis observed that treatment with dexamethasone or vitamin B12 alone, or treatment with both agents, led to a much larger number of Schwann cells and myelinated nerve fibres compared with that in the saline group ($P < 0.05$) [41]. Okada et al. showed in a rat sciatic nerve injury model that continuous administration of high doses of methylcobalamin can improve nerve regeneration and functional recovery [40].

According to Shortland et al., a single dose of 0.1 μM riluzole was sufficient to promote neuronal survival in neonatal dorsal root ganglion cultures, whereas repeated riluzole administration was necessary in adult cultures. For both types of injuries, riluzole enhanced neurite outgrowth (number, length and branch pattern) significantly more on the injured side in comparison with the contralateral side [137].

Tseng et al. found that once-daily administration of 10 μg of 4-aminopyridine enhanced the speed of recovery from crush injury when it was used as early as 3 days post-injury in mice treated daily (beginning 24 hours post-injury), and a significant improvement (>25%) in gait function was confirmed over the control groups (vehicle-treated animals). Furthermore, at 5 and 8 days post-injury, 4-AP-treated mice showed statistically significant twofold greater levels of improvement than the control groups [142].

Han et al. carried out a study in which the right sciatic nerve of adult rats was transected and sutured, and then a gelfoam soaked with verapamil solution for 4 weeks was topically applied by them. The results showed that verapamil can inhibit the secretion of extracellular matrix from fibroblasts *in vivo* through suppression of type I and III collagen secretion. Verapamil also increased the total number of axons as well as the number of myelinated axons more than in the control group, in which gelfoam soaked with physiological saline was topically applied ($P < 0.05$) [44].

12. Conclusions

Generally, it should be mentioned that the most of the experimental research discussed in this chapter was conducted in rats and rabbits, in which the above-mentioned pharmacological agents have been applied (mainly locally in sciatic nerve). The success of the regenerative process of nerve repair in experimental research can be evaluated using a variety of methods, such as morphological, immunohistochemical, electrophysiological, biomechanical and functional evaluation. Some of the pharmacological agents described in this chapter are still only used for experimental purposes, whereas some of them are in clinical use for the treatment of various diseases, and now their neuroregenerative effects will also be explored in experimental animals. However, the success of nerve regeneration depends on the type and degree of nerve injury, age, repair time, operative techniques and the type of materials used. By combining appropriate dosages of these pharmacological agents with improved microsurgical techniques for nerve repair, better experimental results may be achieved in the future, encouraging clinical application of these agents. However, it is understandable that complete regeneration and functional recovery will almost never be achieved, regardless of the operative technique used or the type of pharmacological agent applied.

Author details

Agon Mekaj[1] and Ymer Mekaj[2*]

*Address all correspondence to: ymer.mekaj@uni-pr.edu

1 Clinic of Neurosurgery, Faculty of Medicine, University of Prishtina, Prishtina, Kosovo

2 Institute of Pathophysiology, Faculty of Medicine, University of Prishtina, Prishtina, Kosovo

References

[1] McCance KL, Heuther SE. Pathophysiology: The Biologic Basis for Disease in Adults and Children. 5th ed. Philadelphia, USA: Elsevier Mosby; 2006. p 411.

[2] Geuna S, Tos P, Titolo P, Ciclamini D, Beningo T, Battiston B. Update on nerve repair by biological tubulization. Journal of Brachial Plexus and Peripheral Nerve Injury. 2014;**9**:3. DOI: 10.1186/1749-7221-9-3

[3] Madura T: Pathophysiology of Peripheral Nerve Injury, Basic Principles of Peripheral Nerve Disorders, Dr. Seyed Mansoor Rayegani (Ed.), ISBN: 978-953-51-0407-0, InTech; 2012. Chapter 1, pp 1-16.

[4] Kerfeld CA, Sawaya MR, Tanaka S, Nguyen CV, Phillips M, Beeby M, Yeates TO. Protein structures forming the shell of primitive bacterial organelles. Science. 2005;**309**:936-938. DOI: 10.1126/science.1113397

[5] Martini AI, Frederic ET. Anatomy and Physiology. 1st ed. Philipine: Pearson Education South Asia Pte. Ltd; 2007

[6] Siemionow M, Brzezicki G. Chapter 8: Current techniques and concepts in peripheral nerve repair. International Review of Neurobiology. 2009;**87**:141-172. DOI: 10.1016/S0074-7742(09)87008-6

[7] Kouyoumdjian JA. Peripheral nerve injuries: A retrospective survey of 456 cases. Muscle Nerve. 2006;**34**:785-788. DOI: 10.1002/mus.20624

[8] Eser F, Aktekin LA, Bodur H, Atan C. Etiological factors of traumatic peripheral nerve injuries. Neurology India. 2009;**57**:434-437. DOI: 10.4103/0028-3886.55614

[9] Robinson LR. Traumatic injury to peripheral nerves. Muscle Nerve. 2000;**23**:863-873

[10] Jacques L, Kline DG. Response of the peripheral nerve to physical injury. In: Crockard A, Hayward R, Hoff JT, editors. Neurosurgery: The Scientific Basis of Clinical Practice. 3rd ed, Vol. 1. London: Blackwell; 2000. pp. 516-525

[11] Burnett MG, Zager EL. Pathophysiology of peripheral nerve injury: A brief review. Neurosurgical Focus. 2004;**16**:E1

[12] Zochodne DW, Levy D. Nitirc oxide in damage, disease and repair of the peripheral nervous system. Cellular and Molecular Biology (Nosi-le-grand). 2005;**51**:255-267

[13] Menorca RM, Fussell TS, Elfar JC. Nerve physiology: Mechanisms of injury and recovery. Hand Clinics. 2013;**29**:317-330. DOI: 10.1016/j.hcl.2013.04.002

[14] Seddon HJ. Three types of nerve injury. Brain. 1943;**66**:237-288

[15] Sunderland S. Nerves and Nerve Injuries. 2nd ed. London: Churchill Livingston; 1978. pp 827-828.

[16] Flores AJ, Lavernia CJ, Owens PW. Anatomy and physiology of peripheral nerve injury and repair. American Journal of Orthopedics (Belle Mead NJ). 2000;**29**:167-173

[17] Geuna S, Raimondo S, Ronchi G, Di Spicio F, Tos P, Czaja Fornaro M. Chapter 3: Histology of the peripheral nerve and changes occurring during nerve regeneration. International Review of Neurobiology. 2009;**87**:27-46. DOI: 10.1016/S0074-7742(09)87003-7

[18] Grinsell D, Keating CP. Peripheral nerve reconstruction after injury: A review of clinical and experimental therapies. BioMed Research International. 2014;**2014**:698256. DOI: 10.1155/2014/698256

[19] Pfister BJ, Gordon T, Loverde JR, Kochar AS, Mackinnon SE, Cullen DK. Biomedical engineering strategies for peripheral nerve repair: Surgical applications, state of the art, and future challenges. Critical Reviews in Biomedical Engineering. 2011;**39**:81-124

[20] Waller A. Experiments on the section of the glossopharyngeal and hypoglossal nerves of the frog, and observations of the alterations produced thereby in the structure of their primitive fibres. Philosophical Transactions of the Royal Society of London. 1850;**140**:423-429

[21] Fawcett JW, Keynes RJ. Peripheral nerve regeneration. Annual Review of Neuroscience. 1990;**13**:43-60. DOI: 10.1146/annurev.ne.13.030190.000355

[22] Mira JC. Effects of repeated denervation on muscle reinnervation. Clinics in Plastic Surgery. 1984;**11**:31-38

[23] Jenq CB, Jenq LL, Coggeshall RE. Numerical patterns of axon regeneration that follow sciatic nerve crush in the neonatal rat. Experimental Neurology. 1987;**95**:492-499

[24] Son YJ, Thompson WJ. Schwann cell processes guide regeneration of peripheral axons. Neuron. 1995;**14**:125-132

[25] Mekaj AY, Morina AA, Bytyqi CI, Mekaj YH, Duci SB. Application of topical pharmacological agents at the site of peripheral nerve injury and methods used for evaluating the success of the regenerative process. Journal of Orthopaedic Surgery and Research. 2014;**9**:94. DOI: 10.1186/s13018-014-0094-3

[26] Gold BG. FK506 and the role of immunophilins in nerve regeneration. Molecular Neurobiology. 1997;**15**:285-306. DOI: 10.1007/BF02740664

[27] Choi BH, Zhu SJ, Kim MY, Huh JY, Lee SH, Jung JH. Transplantation of cultured bone marrow stroma cells to improve peripheral nerve regeneration. International Journal of Oral and Maxillofacial Surgery. 2005:**34**:537-542. DOI: 10.1016/j.ijom.2004.10.017

[28] Tung TH. Tacrolimus (FK506): Safety and applications in reconstructive surgery. Hand (New York). 2010;**5**:1-8. DOI: 10.1007/s11552-009-9193-8

[29] Konofaos P, Terzis JK. FK506 and nerve regeneration: Past, present and future. Journal of Reconstructive Microsurgery. 2013;**29**:141-148. DOI: 10.1055/s-0032-1333314

[30] Hiro D, Ito A, Matsuta K, Mori Y. Hialuronic acid is an endogenous induces of interleukin-1 production by human monocytes and rabbit macrophages. Biochemical and Biophysical Research Communications. 1986;**140**:715-722

[31] Ozgenel GY. Effects of hyaluronic acid on the peripheral nerve scarring and regeneration in rats. Microsurgery. 2003;**23**:575-581. DOI: 10.1002/micr.10209

[32] Smit X, van Neck JW, Afoke A, Hovius SE. Reduction of neural adhesions by biodegradable autocrosslinked hyaluronic acid gel after injury of peripheral nerves: An experimental study. Journal of Neurosurgery. 2004;**101**:648-652. DOI: 10.3171/jns. 2004.101.4.0648

[33] Odaci E, Kaplas S. Melatonin and nerve regeneration. Chapter 16. International Review of Neurobiology. 2009;**87**:317-335. DOI: 10.1016/S0074-7742(09)87016-5

[34] Turgut M, Kaplan S. Effects of melatonin on peripheral nerve regeneration. Recent Patents on Endocrine Metabolic & Immune Drug Discovery. 2011;**5**:100-108

[35] Behram Kandemir Y, Sarikcioglu L. Melatonin and its therapeutic actions on peripheral nerve regeneration. Folia Morphologica (Warsz). 2015;**74**:283-289. DOI: 10.5603/FM.2015.0043

[36] Nachemson AK, Lundborg G, Myrhage R, Rank F. Nerve regeneration and pharmacological suppression of the scar reaction at the suture site. An experimental study on the effect of estrogen progesterone, methylprednisolone-acetate and cis-hydroxyproline in rat sciatic nerve. Scandinavian Journal of Plastic and Reconstructive Surgery. 1985;**19**:255-260

[37] Boa F, Chen Y, Dekaban GA, Weaver LC. Early anti-inflammatory treatment reduces lipid peroxidation and protein nitration after spinal cord injury in rats. Journal of neurochemistry. 2004;**88**:1335-1344

[38] Mahammadi R, Azad-Tirgan M, Amini K. Dexamethasone topically accelerates peripheral nerve repair and target organ reinnervation: A transected sciatic nerve model in rat. Injury. 2013;**44**:565-569. DOI: 10.1016/j.injury.2012.10.013

[39] Suslu H, Altun M, Erdivanli B, Turan SH. Comparison of the effect of local and systemic dexamethasone on the rat traumatic sciatic nerve model. Turkish Neurosurgery. 2013;**23**:623-629. DOI: 10.5137/1019-5149.JTN.7761-13.0

[40] Okada K, Tanaka H, Temporin K, Okamoto M, Kuroda Y, Moritomo H, Murase T, Yoshikawa H. Methylcobalamin increases Erk1/2 and Akt activities through the methylation cycle and promotes nerve regeneration in a rat sciatic nerve injury model. Experimental Neurology. 2010;**222**:191-203. DOI: 10.1016/j.expneurol.2009.12.017

[41] Sun H, Yang T, Li Q, Zhu Z, Wang L, Bai G, Li D, Li Q, Wang W. Dexamethasone and vitamin B12 synergistically promote peripheral nerve regeneration in rats by upregulating the expression of brain-derived neurotrophic factor. Archives of Medical Science. 2012;**8**:924-930. DOI: 10.5114/aoms.2012.31623

[42] Altun I, Kurutas EB. Vitamin B complex and vitamin B12 levels after peripheral nerve injury. Neural Regeneration Research. 2016;**11**:842-845. DOI: 10.4103/1673-5374.177150

[43] Darwish Iel S, Dessouky IS. Potential neuroprotective role of verapamil in experimentally-induced chronic sciatic nerve constriction in mice. British Journal of Medicine and Medical Research. 2015;**8**:781-789. DOI: 10.9734/ BJMMR/2015/17908

[44] Han AC, Deng JX, Huang QS, Zheng HY, Zhou P, Liu ZW, Chen ZB. Verapamil inhibits scar formation after peripheral nerve repair *in vivo*. Neural Regeneration Research. 2016;**11**:508-511. DOI: 10.4103/1673-5374.179075

[45] Leinster VH, Robson LG, Shortland PJ. Differential effects of riluzole on subpopulations of adult rat dorsal root ganglion neurons *in vitro*. Neuroscience. 2010;**166**:942-951.DOI: 10.1016/j.neuroscience.2009.12.058

[46] Ghayour MB, Abdolmaleki A, Behnam-Rassouli M. The effect of Riluzole on functional recovery of locomotion in the rat sciatic nerve crush model. European Journal of Trauma and Emergency Surgery. DOI: 10.1007/s00068-016-0691-4

[47] Balazs EA, Laurent TC, Howe AF, Varga L. Irradiation of mucopolysaccharides with ultraviolet light and electrons. Radiation Research. 1959;**ll**:149-164

[48] Laurent TC. Biochemistry of hyaluronan. Acta oto-laryngologica Supplementum 1987; **442**:7-24

[49] Laurent TC, Laurent UBG, Fraser JR. Functions of hyaluronan. Annals of the Rheumatic Diseases. 1995;**54**:429-432

[50] Becker LC, Bergfeld WF, Belsito DV, Klaassen CD, Marks Jr JG, Shank RC, Slaga TJ, Snyder PW. Cosmetic Ingredient Review Expert Panel, Andersen FA. Final report of the safety assessment of hyaluronic acid, potassium hyaluronate, and sodium hyaluronate. International Journal of Toxicology. 2009;**28**:5-67. DOI: 10.1177/1091581809337738

[51] Meyer L, Palmer J. The polysaccharide of the vitreous humor. Biological Chemistry. 1934; **107**:629–634

[52] Balazs EA, Laurent TC, Jeanloz RW. Nomenclature of hyaluronic acid. Biochemical Journal. 1986;**235**:903

[53] Kreil G. Hyaluronidases-A group of neglected enzymes. Protein Science. 1995;**4**:1666-1669. DOI: 10.1002/pro.5560040902

[54] Weigel PH. Hyaluronan synthase: The mechanism of initiation at the reducing end and a pendulum model for polysaccharide translocation to the cell exterior. International Journal of Cell Biology. 2015;**2015**:367579. DOI: 10.1155/2015/367579

[55] DeAngelis PL, Papaconstantinou J, Weigel PH. Isolation of a Streptococcus pyogenes gene locus that directs hyaluronan biosynthesis in acapsular mutants and in heterologous bacteria. Journal of Biological Chemistry. 1993;**268**:14568-14571

[56] Dougherty BA, van de Rijn I. Molecular characterization of hasA from an operon required for hyaluronic acid synthesis in group A streptococci. Journal of Biological Chemistry. 1994;**269**:169-175

[57] DeAngelis PL, Weigel PH. Immunochemical confirmation of the primary structure of streptococcal hyaluronan synthase and synthesis of highmolecular weight product by the recombinant enzyme. Biochemistry. 1994;**33**:9033-9039

[58] Tien JY, Spicer AP. Three vertebrate hyaluronan synthases are expressed during mouse development in distinct spatial and temporal patterns. Developmental Dynamics. 2005;**233**:130-141

[59] Törrönen K, Nikunen K, Kärnä R, Tammi M, Tammi R, Rilla K. Tissue distribution and subcellular localization of hyaluronan synthase isoenzymes. Histochemistry and Cell Biology. 2014;**141**:17-31. DOI: 10.1007/s00418-013-1143-4

[60] Toole BP, Wight TN, Tammi MI. Hyaluronan-cell interactions in cancer and vascular disease. Journal of Biological Chemistry. 2002;**277**:4593–4596

[61] Mekaj Y, Mekaj A. Prevention of failed back surgery syndrome with applications of different pharmacological agents: A review article. Therapeutic Targets for Neurological Diseases. 2015;**2**:e507

[62] Medina JM, Thomas A, Denegar CR. Knee osteoarthritis: Should your patient opt for hyaluronic acid injection? Journal of Family Practice. 2006;**55**(8) 669-675

[63] Necas J, Bartosikova L, Brauner P, Kolar J. Hyaloronic acid (hyaloronan): A review. Veterinarni Medicina. 2008;**52**:397-411

[64] Burd DA, Greco RM, Regauer S, Longaker MT, Siebert JW, Garg HG. Hyaloronan and wound healing: A new perspective. British Journal of Plastic Surgery. 1991;**44**:579-584

[65] Balazs EA, Denlinger JL: The role of hyaluronic acid in arthritis and its therapeutic use. In: Peyron JG, editor. Osteoarthritis: Current Clinical and Fundamental Problems. Geigy: Basel; 1984. pp. 165-174

[66] Amarnath LP, Srinivas A, Ramamurthi A. *In vitro* hemocompatibility testing of UV-modified hyaluronan hydrogels. Biomaterials. 2006;**27**:1416-1424

[67] Chen JM, Lee SH, Tsai TT, Niu CC, Chen LH, Chen WJ. Anti-adhesive effect of hyaluronate in a rabbit laminectomy model. Biomedical Journal 2014;**37**:218-224

[68] Isik S, Taskapilioğlu MÖ, Atalay FO, Dogan S. Effects of cross-linked hight-moleculare-weight hyaluronic acid on epidural fibrosis: Experimental study. Journal of Spine & Neurosurgery. 2015;**22**:94-100. DOI: 10.3171/2014.10.SPINE131147

[69] Adanali G, Verdi M, Tuncel A, Erdogan B, Kargi E. Effect of hyaluronic acid-carboxymethylcellulose membrane on extraneural adhesion formation and peripheral nerve regeneration. Journal of Reconstructive Microsurgery. 2003;**19**:29-36. DOI: 10.1055/s-2003-37188

[70] Park JS, Lee JH, Han CS, Chung DW, Kim GY. Effect of hyaluronic acid-carboxymethylcellulose solution on perineural scar formation after sciatic nerve repair in rats. Clinics in Orthopedic Surgery 2011;**3**: 315-324. DOI: 10.4055/cios.2011.3.4.315

[71] Agenor A, Dvoracek L, Leu A, Hunter DA, Newton P, Yan Y, Johnson PJ, Mackinnon SE, Moore AM, Wood MD: Hyaluronic acid/carboxymethyl cellulose directly applied to transected nerve decreases axonal outgrowth. J Biomed Mater Res B Appl Biomater. 2017;105(3):568-574. DOI: 10.1002/jbm.b.33576

[72] Torigoe K, Tanaka HF, Ohkochi H, Miyasaka M, Yamanokuchi H, Yoshidad K, Yoshida T. Hyaluronan tetrasaccharide promotes regeneration of peripheral nerve: *In vivo* analysis by film model method. Brain Research. 2011;**1385**:87-92. DOI: 10.1016/j.brainres.2011.02.020

[73] Mohammad JA, Warnke PH, Pan YC, Shenaq S: Increased axonal regeneration through a biodegradable amnionic tube nerve conduit: Effect of local delivery and incorporation of nerve growth factor hyaluronic acid media. Annals of Plastic Surgery. 2000;**44**: 59-64

[74] Zor F, Deveci M, Kilic A, Ozdag MF, Kurt B, Sengezer M, Sönmez TT. Effect of VEGF gene therapy and hyaluronic acid film sheath on peripheral nerve regeneration. Microsurgery. 2014;**34**:209–216. DOI: 10.1002/micr.22196.

[75] Yang MH, Chen KC, Chiang PW, Chung TW, Chen WJ, Chu PY, Chen SC, Lu YS, Yuan CH, Wang MC, Lin CY, Huang YF, Jong SB, Lin PC,Tyan YC. Proteomic profiling of neuroblastoma cells adhesion on hyaluronic Acid-Based surface for neural tissue engineering.BioMed Research International. 2016;**2016**:1917394. DOI: 10.1155/2016/1917394

[76] Mekaj AY, Manxhuka-Kerliu S, Morina AA, Duci SB, Shahini L, Mekaj YH. Effects of hyaluronic acid and tacrolimus on the prevention of perineural scar formation and on nerve regeneration after sciatic nerve repair in a rabbit model. European Journal of Trauma and Emergency Surgery. DOI: 10.1007/s00068-016-0683-4.

[77] Wallemacq PE, Reding R. FK506 (tacrolimus), a novel immunosuppressant in organ transplantation: Clinical, biomedical, andanalytical aspects. Clinical Chemistry. 1993;**39**: 2219-2228

[78] Kino T, Hatanaka H, Miyata S, Inamura N, Nishiyama M, Yajima T, Goto T, Okuhara M, Kohsaka M, Aoki H, Ochiai T. FK-506, a novel iinmunosuppreasant isolated from a streptomyce8. II. Immunosuppressive effect of FK-506 *in vitro*. Journal of Antibiotics (Tokyo). 1987;**40**:1256-1265. DOI.org/10.7164/antibiotics.40.1256

[79] Sawada S, Suzuki G, Kawase Y, Takaku F. Novel immunosuppressive agent, FK506. *In vitro* effects on the doned T cell activation. Journal of Immunology. 1987;**139**:1797-1803

[80] Goto T, Kino T, Hatanaka H, Nishiyama M, Okuhara M, Kohsaka M, Aoki H, Imanaka H. Discovery of FK 506, a novel immunosuppreasant isolated from Streptomycea tsukubaensis. Transplantation Proceedings. 1987;**19**:4-8

[81] Ochiai T, Nakima K, Nagata M, Suzuki T, Asano T, Uematsu T, Goto T, Hori S, Kenmochi T, Nakagoori T. Effect of a new immunosuppressive agent, FK 506, on heterotopic cardiac allo. transplantation in the rat. Transplantation Proceedings. 1987;**19**:1284-1286

[82] Pleyer U, Lutz S, Jusko WJ, Nguyen KD, Narawane M, Ruckert D, Mondino BJ, Lee VH, Nguyen K. Ocular absorption of topically applied FK506 from liposomal and oil formulations in the rabbit eye. Investigative Ophthalmology & Visual Science. 1993;**34**:2737-2742

[83] Tung TH. Clinical strategies to enhance nerve regeneration. Neural Regeneration Research. 2015;**10**:22-24. DOI: 10.4103/1673-5374.150641

[84] Yeh C, Browers D, Hadlock TA. Effect of FK506 on functional recovery after facial nerve injury in the rat. Archives of Facial Plastic Surgery. 2007;**9**:333-339. DOI: 10.1001/archfaci.9.5.333

[85] Toll EC, Seifalian AM, Birchall MA. The role of immunophilin ligands in nerve regeneration. Regenerative Medicine. 2011;**6**:635-652. DOI: 10.2217/rme.11.43

[86] Saganová K, Gálik J, Blaško J, Korimová A, Račeková E, Vanický I. Immunosuppressant FK506: Focusing on neuroprotective effects following brain and spinal cord injury. Life Sciences. 2012;**91**:77-82. DOI: 10.1016/j.lfs.2012.06.022

[87] Yan L, Li X, Wang J, Sun Y, Wang D, Gu J, He J, Hu H, Chen G, Wang Q, Feng X. Immunomodulatory effectiveness of tacrolimus in preventing epidural scar adhesion after laminectomy in rat model. European Journal of Pharmacology. 2013;**699**:194-199. DOI: 10.1016/j.ejphar.2012.11.037

[88] Bavetta S, Hamlyn PJ, Burnstock G, Lieberman AR, Anderson PN. The effects of FK506 on dorsal column axons following spinal cord injury in adult rats: Neuroprotection and local regeneration. Experimental Neurology.1999;**158**:382–393. DOI: 10.1006/exnr.1999.7119.

[89] Arun A, Abt NB, Tuffaha S, Brandacher G, Leto Barone AA. Nerve regeneration in vascularized composite allotransplantation: Current strategies and future directions. Plastic and Aesthetic Research. 2015;**2**:226-235. DOI:10.4103/2347-9264.158853.

[90] Yan Y, Sun HH, Hunter DA, Mackinnon SE, Johnson PJ. Effect of short-term FK506 administration on accelerating nerve regeneration. Neurorehabilitation and Neural Repair. 2012;**26**: 570-580. DOI: 10.1177/1545968311431965

[91] Azizi S, Mohammadi R, Amini K, Fallah R. Effects of topically administered FK506 on sciatic nerve regeneration and reinnervation after vein graft repair of short nerve gaps. Neurosurgical Focus. 2012, **32**:E5-10. DOI: 10.3171/2012.1.FOCUS11320.

[92] Dawson TM, Steiner JP, Dawson VL, Dinerman JL, Uhl GR, Synder SH. Immunosuppressant FK506 enhances phosphorylation of nitric oxide synthetase and protects against glutamate neurotoxicity. Proceedings of the National Academy of Sciences of the United States. 1993;**90**: 9808-9812

[93] Madsen JR, MacDonald P, Irwin N, Goldberg DE, Yao GL, Meiri KF, Rimm IJ, Stieg PE. Tacrolimus (FK506) increase neuronal expression of GAP-43 and improves functional recovery after spinal cord injury in rats. Experimental Neurology. 1998;**154**: 673-683. DOI: 10.1006/exnr.1998.6974

[94] Que J, Cao Q, Sui T, Du S, Kong D, Cao X. Effect of FK506 in reducing scar formation by inducing fibroblast apoptosis after sciatic nerve injury in rats. Cell Death & Disease. 2013;**4**:e526-e510. DOI: 10.1038/cddis.2013.56

[95] Behram Kandemir Y, Sarikcioglu L. Immunophilin ligands in peripheral nerve regeneration. Folia Morphol (Warsz). 2015;**74**:407-413. DOI: 10.5603/FM.2015.0100

[96] Shahraki M, Mohammadi R, Najafpour A. Influence of tacrolimus (FK506) on nerve regeneration using allografts: A rat sciatic nerve model. Journal of Oral and Maxillofacial Surgery. 2015;**73**:1438 e1–1438 e 9. DOI: 10.1016/j.joms.2015.03.032

[97] Fansa H, Keilhoff G, Altmann S, Plogmeier K, Wolf G, Schneider W. The effect of the immunosuppressant FK 506 on peripheral nerve regeneration following nerve grafting. Journal of Hand Surgery British. 1999;**24**:38-42

[98] Mekaj AY, Morina AA, Manxhuka-Kerliu S, Neziri B, Duci SB, Kukaj V, Miftari I. Electrophysiological and functional evaluation of peroneal nerve regeneration in rabbit following topical hyaluronic acid or tacrolimus application after nerve repair. Nigerian Postgraduate Medical Journal. 2015;**22**:179–184. DOI: 10.4103/1117-1936.170738

[99] Mekaj AY, Morina AA, Lajqi S, Manxhuka-Kerliu S, Kelmendi FM, Duci SB. Biomechanical properties of the sciatic nerve following repair: Effects of topical application of hyaluronic acid or tacrolimus. International Journal of Clinical and Experimental Medicine. 2015;**8**:20218–20226

[100] Watts R, Clunie G, Hall F, Marshall T. Rheumatology.1 edition. New York, USA. Oxford University Press: 2009. p. 558. ISBN 9780199229994.

[101] Borel JF, Feurer C, Gubler HU, Stahelin H. Biological effects of cyclosporin A: A new antilymphocytic agent. Agents Actions. 1976;**6**:468–475

[102] Çetinalp NE, Albayrak SB, Ismailoğlu Ö, Temiz NÇ, Solmaz İ, Tanriöver G, Demir N: Topical Application of Cyclosporine Reduces Epineurial Fibrosis: Gross Postsurgical, Histopathological and Ultrastructural Analysis in a Rat Sciatic Nerve Model. Turk Neurosurg. 2016 [Epub ahead of print]. DOI: 10.5137/1019-5149.JTN.17871-16.1.

[103] Meirer R1, Babuccu O, Unsal M, Nair DR, Gurunluoglu R, Skugor B, Meirer B, Siemionow M. Effect of chronic cyclosporine administration on peripheral nerve regeneration: A dose–response study. Annals of Plastic Surgery. 2002;**49**:96-103

[104] Erkutlu I, Alptekin M, Geyik S, Geyik AM, Gezgin I, Gök A. Early cyclosporin A treatment retards axonal degeneration in an experimental peripheral nerve injection injury model. Neural Regeneration Research. 2015;**10**:266-270. DOI: 10.4103/1673-5374.152381

[105] Matsuda S, Koyasu S. Mechanisms of action of cyclosporine. Immunopharmacology. 2000;**47**:119-125

[106] Sun L, Youn HD, Loh C, Stolow M, He W, Liu JO. Cabin 1, a negative regulator for calcineurin signaling in T lymphocytes. Immunity. 1998;**8**:703-711

[107] Bain JR, Mackinnon SE, Hudson AR, Falk RE, Falk JA, Hunter DA, Makino A. Preliminary report of peripheral nerve allografting in primates immunosuppressed with cyclosporine. Transplantation Proceedings. 1989;**21**:3176-3177

[108] Muramatsu K, Doi K, Kawai S. Nerve regenerating effect of short-course administration of cyclosporine after fresh peripheral nerve allotransplantation in the rat: Comparison of nerve regeneration using different forms of donor nerve allografts. Microsurgery. 1995;**16**:496-504

[109] Hellenbrand DJ, Kaeppler KE, Ehlers ME, Thompson CD, Zurko JC, Buchholz MM, Springer AR, Thompson DL, Ibrahim RK, Hanna A. Immunohistochemical assessment of rat nerve isografts and immunosuppressed allografts. Neurology Research. 2016;**38**:1094-1101

[110] Mohammadi R, Shariati H, Sharghi N, Naghipour M. Peripheral nerve regeneration using silicone rubber chamber combined by local administration of cyclosporin a in streptozotocin induced diabetic rats. International Journal of Oral and Craniofacial Science. 2015;**1**:019-026

[111] Wolrath L. Functional anatomy of the human pineal gland. In: Reiter RJ, editor. The Pineal Gland. New York: Raven Press; 1984. pp. 285-322

[112] Kaya Y, Sarıkcıoğlu L, Aslan M, Kencebay C, Demir N, Derin N, Angelov DN, Yıldırım FB. Comparison of the beneficial effect of melatonin on recovery after cut and crush sciatic nerve injury: A combined study using functional, electrophysiological, biochemical, and electron microscopic analyses. Child's Nervous System. 2013;**29**:389-401. DOI: 10.1007/s00381-012-1936-0

[113] Chang HM, Huang YL, Lan CT, Wu UI, Hu ME, Youn SC. Melatonin preserves superoxide dismutase activity in hypoglossal motoneurons of adult rats following peripheral nerve injury. Journal of Pineal Research. 2008:**44**:172-180. DOI: 10.1111/j.1600-079X.2007.00505.x

[114] Hardeland R, Coto-Montes A, Poeggeler B. Circadian rhythms, oxidative stress, and antioxidative defense mechanisms. Chronobiology International. 2003;**20**:921-962

[115] Chang HM, Ling EA, Lue JH, Wen CY, Shieh JY. Melatonin attenuates neuronal NADPH-d/NOS expression in the hypoglossal nucleus of adult rats following peripheral nerve injury. Brain Research. 2000;**873**:243-251

[116] Abdel-Wanis ME, Kawahara N. The role of neurofibromin and melatonin in pathogenesis of pseudoarthrosis after spinal fusion for neurofibromatous scoliosis. Medical Hypotheses. 2002;**58**:395-398. DOI: 10.1054/mehy.2001.1467

[117] Chang HM, Liu CH, Hsu WM, Chen LY, Wang HP, Wu TH, Chen KY, Ho WH, Liao WC. Proliferative effects of melatonin on Schwann cells: Implication for nerve regeneration following peripheral nerve injury. Journal of Pineal Research. 2014;**56**:322–332. DOI: 10.1111/jpi.12125

[118] Turgut M, Uyanikgil Y, Baka M, Tunç AT, Yavasoglu A, Yourtseven ME, Kaplan S. Pinealectomy exaggerates and melatonin treatment suppresses neuroma formation of transected sciatic nerve in rats. Journal of Pineal Research. 2005;**38**:284-291. DOI: 10.1111/j.1600-079X.2004.00205.x

[119] Kaya Y, Savas K, Sarikcioglu L, Yaras N, Angelov DN. Melatonin leads to axonal regeneration, reduction in oxidative stress, and improved functional recovery following sciatic nerve injury. Current Neurovascular Research. 2015;**12**:53-62

[120] Yanilmaz M, Akduman D, Sagun ÖF, Haksever M, Yazicilar O, Orhan I, Akpolat N, Gök U. The effects of aminoguanidine, methylprednisolone, and melatonin on nerve recovery in peripheral facial nerve neurorrhaphy. Journal of Craniofacial Surgery. 2015;**26**:667-672. DOI: 10.1097/SCS.0000000000001503

[121] Li H, Zhang L, Xu M. Dexamethasone prevents vascular damage in early-stage nonfreezing cold injury of the sciatic nerve. Neural Regeneration Research. 2016;**11**:163-167. DOI: 10.4103/1673-5374.175064

[122] Regina G, Angiletta D, Impedovo G, Robertis GD, Fiorella M, Carratu MR. Dexamethasone minimizes the risk of cranial nerve injury during CEA. Journal of Vascular Surgery. 2009;**49**:99-102. DOI: 10.1016/j.jvs.2008.08.049

[123] Feng X, Yuan W. Dexamethasone enhanced functional recovery after sciatic nerve crush injury in rats. Biomed Research International. 2015:627923. DOI: 10.1155/2015/627923

[124] Hall ED. The neuroprotective pharmacology of methylprednisolone. Journal of Neurosurgery. 1992;**76**:13-22. DOI: 10.3171/jns.1992.76.1.0013

[125] Sadraie SH, Parivar K, Arabi F, Moattari M, Kaka G, Mansouri K. Study of transected sciatic nerve repair by amniotic membrane with betamethasone in adult albino wistar rats. Archives of Iranian Medicine. 2016;**19:**612-617. DOI: 0161909/AIM.003

[126] Li Q, Li T, Cao XC, Luo DQ, Lian KJ. Methylprednisolone microsphere sustained-release membrane inhibits scar formation at the site of peripheral nerve lesion. Neural Regeneration Research. 2016;**11**:835-841. DOI: 10.4103/1673-5374.182713.

[127] McDowell LR. Vitamins in Animal and Human Nutrition, Iowa State University Pres. USA; 2008.

[128] Goyal V. Vitamin B12—A brief overview. IAIM. 2015;**2**:211-216

[129] Wang SA, Yang J, Zhang GB, Feng YH, Wang F, Zhou PY. Effect of mecobalamin treatment on the recovery of patients with posterior communicating artery aneurysm inducing oculomotor nerve palsy after operation. European Review for Medical and Pharmacological Sciences. 2015;**19**:2603-2607

[130] Jolivalt CG, Mizisin LM, Nelson A, Cunha JM, Ramos KM, Bonke D, Calcutt NA. B vitamins alleviate indices of neuropathic pain in diabetic rats. European Journal of Pharmacology. 2009;**612**:41-47. DOI: 10.1016/j.ejphar.2009.04.028

[131] Yamada K. Chapter 9. Cobalt: Its role in health and disease. In: Sigel A, Sigel H, Sigel Roland KO, editors. Interrelations Between Essential Metal Ions and Human Diseases.Metal Ions in Life Sciences Netherlands: Springer; 2013 pp. 295-320. DOI: 10.1007/978-94-007-7500-8_9

[132] Lopatina T, Kalinina N, Karagyaur M, Stambolsky D, Rubina K, Revischin A, Pavlova G, Parfyonova Y, Tkachuk V. Adipose-derived stem cells stimulate regeneration of peripheral nerves: BDNF secreted by these cells promotes nerve healing and axon growth de novo. PLoS One. 2011;**6**:e17899. DOI: 10.1371/journal.pone.0017899

[133] Doble A. The pharmacology and mechanism of action of riluzole. Neurology. 1996;**47**(6 Suppl 4):S233-S241

[134] Wu Y, Satkunendrarajah K, Teng Y, Chow DS, Buttigieg J, Fehlings MG. Delayed post-injury administration of riluzole is neuroprotective in a preclinical rodent model of cervical spinal cord injury. Journal of Neurotrauma. 2013;**30**:441-452. DOI: 10.1089/neu.2012.2622.

[135] Satkunendrarajah K, Nassiri F, Karadimas SK, Lip A, Yao G, Fehlings MG. Riluzole promotes motor and respiratory recovery associated with enhanced neuronal survival and function following high cervical spinal hemisection. Experimental Neurology. 2016;**276**:59-71. DOI: 10.1016/j.expneurol.2015.09.011

[136] Fehlings MG, Wilson JR, Karadimas SK, Arnold PM, Kopjar B. Clinical evaluation of a neuroprotective drug in patients with cervical spondylotic myelopathy undergoing surgical treatment: Design and rationale for the CSM-protect trial. Spine (Philadelphia Pa 1976). 2013;**38**:S68-S75. DOI: 10.1097/BRS.0b013e3182a7e9b0.

[137] Shortland PJ, Leinster VH, White W, Robson LG. Riluzole promotes cell survival and neurite outgrowth in rat sensory neurones *in vitro*. European Journal of Neuroscience. 2006;**24**:3343-3353. DOI: 10.1111/j.1460-9568.2006.05218.x

[138] Nicholson KJ, Zhang S, Gilliland TM, Winkelstein BA. Riluzole effects on behavioral sensitivity and the development of axonal damage and spinal modifications that occur after painful nerve root compression. Journal of Neurosurgery-Spine. 2014;**20**:751-762. DOI: 10.3171/2014.2.SPINE13672

[139] Solari A, Uitdehaag B, Giuliani G, Pucci E, Taus C. Aminopyridines for symptomatic treatment in multiple sclerosis. Cochrane Database of Systematic Reviews. 2001;(4):CD001330. DOI: 10.1002/14651858.CD001330.

[140] Smith C, Kongsamut S, Wang H, Ji J, Kang J, Rampe D. *In Vitro* electrophysiological activity of nerispirdine, a novel 4-aminopyridine derivative. Clinical and Experimental Pharmacology and Physiology. 2009;**36**:1104-1109. DOI: 10.1111/j.1440-1681.2009.05200.x

[141] Hayes KC. The use of 4-aminopyridine (fampridine) in demyelinating disorders. CNS Drug Reviews. 2004;**10**:295-316

[142] Tseng KC, Li H, Clark A, Sundem L, Zuscik M, Noble M, Elfar J. 4-Aminopyridine promotes functional recovery and remyelination in acute peripheral nerve injury. EMBO Molecular Medicine. 2016;**8**:1409-1420. DOI: 10.15252/emmm.201506035

[143] Lundborg G. A 25-year perspective of peripheral nerve surgery. evolving neuroscientific concepts and clinical significance. Journal of Hand Surgery (American Volume). 2000;**25**:391-414. DOI: 10.1053/jhsu.2000.4165

[144] Lee RC, Ping JA. Calcium antagonists retard extracellular matrix production in connective tissue equivalent. Journal of Surgical Research. 1990;**49**:463-466

[145] Ikeda K, Yamauchi D, Osamura N, Hagiwara N, Tomita K. Hyaluronic acid prevents peripheral nerve adhesion. British Journal of Plastic Surgery. 2003;**56**:341-347

[146] Li X, Wang W, Wei G, Wang G, Zhang W, Ma X. Immunophilin FK506 loaded in chitosan guide promotes peripheralnerve regeneration. Biotechnology Letters. 2010;**32**: 1333-1337. DOI: 10.1007/s10529-010-0287-8

Chapter 7

Adult and Reparative Neurogenesis in Fish Brain

Evgeniya V. Pushchina, Anatoly A. Varaksin,

Mariya E. Stukaneva and Eva I. Zharikova

Additional information is available at the end of the chapter

http://dx.doi.org/10.5772/67951

Abstract

The fish brain has a unique feature of vertebrates—it grows with the growth of body over a lifetime. In this regard, fishes are a convenient model for the study of embryonic and postembryonic development of the central nervous system and of the influence of different factors on these processes. Currently, the mechanisms of adult brain morphogenesis of fish, which retain larval stage for a long time, are poorly understood. This is particularly true for participation of radial glia during morphogenesis of the brain, as well as the presence and distribution of the proliferative zone in the adult fish brain. Another interesting and little known aspect is the posttraumatic ability of fish to form active neurogenic niches. Investigation of the structural organizations of neurogenic niches and special conditions of the extracellular environment, as well as the interactions between neighboring cells in a neurogenic niche, is interesting and relevant direction in the study of the neuronal stem cells biology. Injury of fish brain creates special conditions for the implementation of genetic programs aimed at strengthening the proliferation of progenitor cells, as well as the activation and proliferation activity in the neuronal stem cells.

Keywords: adult neurogenesis, neurogenic niche, radial glia, reparative neurogenesis, proliferation, migration, neuroal differentiation, teleost fishes, regeneration, matrix areas of brain, apoptosis, neuroprotective factors, neural stem cells

1. Introduction

Among vertebrates, fishes are known to be able to effectively restore the structure of cells and fibers after damage of the central nervous system (CNS). They have the ability to restore the number of damaged cells by production of new cells in the matrix areas of the brain and neurogenic niches and the ability to restore the structure of damaged axons of neurons in the spinal cord pathways [1]. However, it is currently unknown how this process is related to the neurogenesis in the adult brain and what elements of the matrix areas of the brain are involved in the reparative neurogenesis in fish. The evolutionarily ancient animal groups are often used as a convenient model for neurogenic studies in adults. The brain of such animals has a large number of periventricular proliferative zones and active zones of secondary neurogenesis [2, 3]. In contrast to the mammalian brain, numerous proliferative regions have been found in adult fish. The presence of such regions was described in *Apteronotus leptorhynchus* [4], *Sparus aurata* [5], *Gasterosteus aculeatus* [6], *Danio* sp. [7, 8], and *Austrolebias* sp. [9].

The regenerative processes in the brain of fish after the damaging impact are determined by a number of factors, which distinguish the dynamics of this process from that in other vertebrates, particularly mammals and humans [10, 11]. It is known that the brain injury in the mammalian brain results in a number of pathological changes associated with the development of an inflammatory response to the toxic effects of glutamate and other inflammatory mediators, and further pathological changes associated with processes of secondary inflammation and involve massive cell death [12, 13]. As a result of CNS trauma, the mammalian cells are exposed to severe necrosis and only a small part of them is eliminated via apoptosis [14]. In the fish brain, the cellular response to the trauma develops in a different scenario. Apoptosis is observed 5 min after the injury, which progresses in the next few days [7]. The elimination of damaged cells is carried out by phagocytes (microglia/macrophages), which remove damaged cells very effectively and provide a «clean» cell death without the remaining damaged cellular material and the development of secondary inflammation [15]. The replacement of the large amounts of dead cells resulted from the damage in the fish brain appears from various sources: the radial glia, centers of primary and secondary proliferation, and neurogenic zones. The high regenerative potential in the central nervous system of fish is provided by the activation of specific regenerative factors [1] and the effect of neuroprotective factors protecting damaged cells and providing long-term survival of cells formed as a result of reparative neurogenesis.

2. Adult neurogenesis and neural regeneration in fish brain

Neuroregenerative properties were investigated in various parts of the fish brain: retina [16], optical tectum [17], spinal cord [11], and cerebellum [17]. According to the «Lesion paradigm» formulated by Zupanc and his colleagues [10], a high regenerative potential of the central nervous system of fish is determined by a number of different processes, including the response of the central nervous system after a damaging effect.

The first few reports related to the development of this theory have been derived from studies on European carp *Carassius vulgaris* after injury of the spinal cord at the cervical/thoracic levels and monitoring of structural recovery within 2 weeks from the date of damage. In these studies, however, the histological studies of crossed pieces of the spinal cord have not been conducted, but the conclusions were made for the first time on the functional repair and restoration of motor activity (ability to swim) after application of the damaging effects [11, 18].

Significant progress in understanding the basic signs of successful regeneration after amputation of the caudal portion of the spinal cord has been made in research on electric fish *A. leptorhynchus* [2]. In animal studies, it was demonstrated by the successful restoration of the lost fragments of the spinal cord, along with the amputated part of the caudal fin. One of the initial stages of the repair process, resulting from an injury is the rapid destruction of damaged cells via apoptosis. The first cells with signs of apoptosis occur in an area of damage within 5 min after the injury, and then the number of cells gradually increases, reaching a maximum value within a few hours.

On the second day, the number of cells gradually declined, reaching the background level after approximately 3 weeks. During this period, only some cells underwent necrosis. An elimination of damaged cells by apoptosis in the brain of fish differs significantly from that of mammals [15]. In contrast, the main process of elimination of damaged cells to the injured area of mammals is necrosis [19]. Apoptosis also affects the small part of cells in the areas surrounding injury. The prevalence of necrosis in the mammalian brain after injury is one of the causes of subsequent secondary inflammation in the lesion [14], which in turn causes a further increase in response of necrotic injuries, resulting in the formation of larger cavities deprived of cells. These cavities are usually restricted area of reactive astrocytes, creating both mechanical and biochemical barriers that impede the growth of nerve fibers and cell migration into the damaged area. Unlike necrosis, apoptotic cells characteristically show overall compression, condensation of the nucleus, and the formation of vesicles, which are subsequently destroyed by the macrophages/microglia [20].

Initially, the numbers of phagocytes in the area of damage were small, but after about 3 days of injury, the number of macrophages begins to increase in the area of injury and in the adjacent areas [21]. The main side effects of necrosis, associated with inflammation of the surrounding tissue, are completely absent in apoptotic "clean" method of elimination of cells. Thus, the prevalence of processes of "clean" cell death for the destruction of damaged cells is a key feature underlying the regenerative capacity of the adult fish brain.

An important aspect of promoting successful regeneration in the brain of the fish is the detection of specific neuroprotective factors which play a key role in maintaining the viability of neurons in the affected areas and prevent further cell death after injury [22]. Such factors are being considered as different substances, in particular some of the calcium binding proteins, such as calbindin-28 and parvalbumin. Expression of these calcium-binding proteins in the cells briefly increases in granular layer of the cerebellum of *A. leptorhynchus* between 16 h and 7 days after injury [23]. It is assumed that the calcium-binding proteins have a protective effect by the buffering of free calcium, the level of which increases considerably after injury. Another neuroprotective factor is the enzyme glutamine synthetase (GS), which conversed synaptically released glutamate in the neutral glutamine. It is known that as a result of damage to the brain cells, the extracellular medium receives a large amount of glutamate, creating hyperexcitation of glutamate receptors and the excitotoxicity [12]. To dispose of glutamate, there arises a necessity of a sufficient amount of the enzyme glutamine synthetase excreted by astroglial cells, carrying out the reuptake of glutamate and converting it into glutamine. According to studies, after traumatic injury of fish, levels of glutamine synthetase significantly increased [24], whereas in the mammalian brain, conversely, decreased [13, 25]. Increased synthesis of glutamine synthetase in the brain of the fish is likely to provide an important mechanism for reducing the neurodegenerative process caused by neurotoxic effects of glutamate. Such differences in the expression of GS in fish's brain and in mammals are certainly interesting because they determine significant limitations of regenerative activity of the brain tissue of mammals in comparison to the fish brain.

3. Apoptosis and cell migration after injury of cerebellum

We observed the apoptosis and migration of cells in the young masou salmon *Oncorhynchus masou* after mechanical injury of the cerebellum. Two days after injury in adjacent zone, we detected significant change in cell composition in molecular and in the granular layers [26]. The most characteristic phenomenon was the emergence of large areas of cell migration from the area of regional neurogenic niches and the largest area of the secondary neurogenesis, located in the dorsomedial part of cerebellar body (**Figure 1A**).

We believe that it was mainly due to the migration process in neurogenic niches and dorsomedial area. The highest density of cells was detected in the vicinity of the puncture area, gradually decreasing with the distance from the area of injury. Near the area of injury, many TUNEL-labeled components corresponding to different stages of the apoptotic process were localized (**Figure 1B**). So, dense apoptotic bodies, which are the final stage of coarse chromatin condensation and apoptotic cell degradation, were found. The size of apoptotic bodies was about 8–10 μm. In areas of apoptotic fragments localization were found large cells with basophilic cytoplasm, the diameter of cells body is about 13 μm. These cells tend to have an irregular shape and had cytoplasmic outgrowths. Presumably, these structures correspond to regional microglia/macrophages involved in phagocytosis of apoptotic fragments and recycling. Along with individual elements, there also occurred small conglomerates, including up to three apoptotic bodies.

Another variety of apoptotic bodies were small TUNEL-labeled bodies representing degranulated fragments of damaged cells. In the most superficial parts of the molecular layer was observed a very large number of small cells lacking the morphological features of differentiation (**Figure 1C**). Such morphological pattern of surface of the molecular layer apparently reflects the intensity of the processes of cell migration from the superficial regions of the cerebellum to the zone of injury. The surface area which has been characterized by a high density of cells revealed TUNEL-labeled small elements corresponding to cell degranulation products in the area.

In our studies, some effects of the damaging of cerebellum were combined with complex morphogenetic background of the ongoing postembryonic development of the brain *O. masou*. The experimental fish was in the process of active growth, resulting in increased proliferative activity in the cerebellum and morphogenetic zones of periventricular regions of the brain (**Figure 1C**). Previously in experiments with *Danio rerio*, it has been found that damage to the

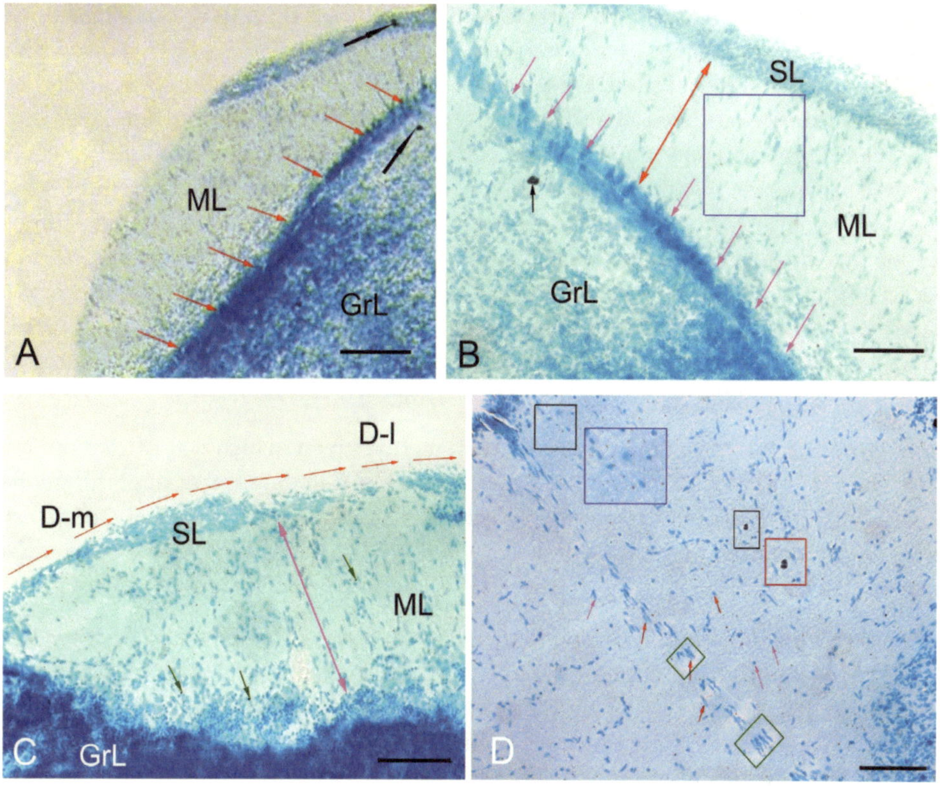

Figure 1. Patterns of cell migration and apoptosis in the cerebellum of juvenile masou salmon *Oncorhynchus masou* after injury. A—common view of the areas of trauma, small arrow shows the area of puncture, black arrows (here and below) show apoptotic bodies; B—dorsomedial part of *corpus cerebellum*, which contains migrating from the surface layer (SL) cells (contoured by rectangle), the big arrow indicates the direction of radial migration; C—lateral part of the molecular layer (ML) contained cells migrating to the area of injury, arrows show the direction of the tangential migration, big arrow—radial migration, dark arrows show TUNEL-labeling fragments of degranulated cells, D-m—dorsomedial, D-l—dorsolateral area; D—the area of median suture, small squares delineated clusters of migrating cells, in other squares are small TUNEL-labeled fragments, and apoptotic cells, arrows show the different types of cells; ML—molecular layer, GrL—granular layer. Scale bar: A—200 μm, B-D—100 μm.

cerebellum causes increased proliferation of cells in the outer regions adjacent to the meninges and also in *valvula cerebelli* and *granular eminentia* [27]. These data were obtained using experimental labeling BrdU of dividing cells [28] and other markers of cell proliferation [29]. High neurogenetic activity after traumatic injury was detected in the cerebellum of *A. leptorhynchus*. In this species belonging to the group of specialized electrical gymnotiformn fish cerebellum, it takes up 75% of total brain volume; it is the largest center for neurogenesis both during normal adult development and in terms of traumatic impact [30].

In juvenile *O. masou* was observed very high initial intensity of proliferation in the primary matrix region of brain (periventricular zone) and in zones of secondary proliferative activity detected in adult animals [31]. The corresponding data obtained by labeling both the proliferative cell nuclear antigen (PCNA) and using traditional morphological methods to assess the mitosis in matrix areas of brain (in particular, the dorsomedial region). The intensity of cell proliferation in this region has been previously described in adults *D. rerio* [7, 32] and *A. leptorhynchus* [4] and juveniles of trout, *Salmo gairdneri* [33]. In our observations, particularly high proliferative activity has been identified in the dorsomedial area, the surface layer and the body of cerebellum, *granular eminentia* and vestibule-lateral areas of a damaged cerebellum. Thus, we can conclude that the proliferation of cells in these areas of the cerebellum, which is already very intense during normal morphogenetic activity within a given period of ontogenesis, is further enhanced after the damaging effects. The presence of newly formed cells in the superficial layer of the cerebellum associates partly with background morphogenetic activity and partly with damaging effects. As a result of damage to the cerebellum in young *O. masou*, enhanced proliferative activity was induced in both traditional areas of adult neurogenesis and the surface layer.

Counting the number of cell nuclei stained with methyl green, which was carried out on the surface layer of the cerebellum, indicates the presence of a large number of undifferentiated cells with high nuclear-cytoplasmic ratio. Such cells based on morphological criteria can be referred as proliferating population and cell population at the early stages of differentiation and/or migration.

We believe that the proliferative response to damage the cerebellum in young *O. masou* should be interpreted taking into account the relatively high background level of activity of the matrix areas of brain [34]. The process of apoptotic cell death accompanies the "normal" adult neurogenesis [35], and, at the same time, it is a physiological response of the nervous system to injury of *O. masou* cerebellum. Apoptosis during normal development has been described in the brain of *A. leptorhynchus* [27]. In these studies, it was found that during the proliferative activity of matrix areas, cerebellar cells formed with signs of somatic aneuploidy. This material is obviously defective, because the relevant units do not have a normal diploid number of chromosomes, and is subject to elimination of apoptotic scenario [36]. In studies on intact adult specimens of Amur sturgeon have been found high values of apoptotic index in different parts of the central nervous system, including the integrative centers of brain (*optic tectum, cerebellum*) and sensory centers of the brain stem (nucleus V and VII cranial nerves pairs) [35]. Similar phenomena are typical for continuing morphogenetic activity in the various centers of the brain of fish, where continued replenishment of new cells occurs throughout life. The resulting cells appeared *de novo* can be integrated into existing neural networks not only during embryonic neurogenesis but also in adult animals. This phenomenon, in particular the special characteristic of the sensory areas, updated with new structural elements as the growth of the animal. Apoptosis in these physiologically active developing systems may play a role of physiological filter that regulates the number of new cells and ensures elimination of "old cells".

Large TUNEL-labeled bodies (**Figure 1D**) conform to the final stage of chromatin degradation in apoptotic cells. This stage is characterized by the formation of large condensed fragments of chromatin that cannot be disposed by macrophages/microglia. These apoptotic bodies were also identified in the morphogenetic studies in mammals [37]. Other visible TUNEL-labeled elements are small weakly diffused particles, which are products of degranulation of cells that are eliminated by apoptosis. These "remnants" of the cells were identified over large areas, located in different parts of the cerebellum. These different types of TUNEL-labeled structures in the cerebellum of *O. masou* were identified almost everywhere. Apoptotic index values that vary significantly in different parts of the cerebellum show different intensity of apoptosis in the matrix zones, areas of trauma, and in adjacent areas of intact regions of the cerebellum of young *O. masou*.

In our studies on the second day from the date of injury to the cerebellum, intensity of TUNEL-labeling of apoptotic bodies was not very high [26]. However, apoptosis in a zone adjacent to the areas of trauma has been well defined. This surely indicates a part of the mechanism in the process of disposing of damaged cells. In the mentioned period of time (2 days after injury) in the area of damage was revealed increased density of distribution undifferentiated cells. This fact indicates that the reparative process moved at a later stage. Apparently, the reparative processes of neurogenesis in juvenile salmonids implemented in earlier periods compared with what is commonly referred to in the literature [1, 27, 28]. This is likely due to the high intensity of the background morphogenetic activity in the cerebellum of young fishes as compared to that in adult animals.

After mechanical trauma of the cerebellum, patterns of tangential and radial cell migration can be observed. The zones of cell migration are best expressed in the dorsal part of cerebellar body, as well as in the areas of secondary neurogenesis. In our experimental conditions, two groups of TUNEL-labeled structures were identified: large TUNEL-labeled bodies, corresponding to the final stage of degradation of apoptotic cells, and small, weakly condensed particles, which are apparently products of cells degranulation. In the matrix areas, areas of trauma, and intact areas of the cerebellum, different levels of apoptotic activity were observed. The highest value of apoptotic index (5%) after the traumatic impact on the cerebellum was observed in the molecular layer, which is the main area of radial migration of cells. Thus, the background morphogenetic processes and physiological repair processes dominate in the cerebellum of young *O. masou* after traumatic exposure. The intensity of apoptosis vary between different areas of masu salmon cerebellum, as these areas differ considerably.

4. *In vivo* investigation of cell migration after mechanical injury

Microglia/macrophages have been identified within a few days after lesions in several divisions of the CNS of teleost fish—the cerebellum [38], the dorsal telencephalon [39, 40], and the retina [41]. We used multiphoton confocal microscopy for the *in vivo* study of early response of microglia/macrophages in the damaged midbrain of juvenile chum salmon *Oncorhynchus keta* [42]. The results obtained allow the use of injection of DiI in the area of brain injury as a method to identify a population of phagocytic cells in the brain, based on the physiological response of macrophages/microglia. Thus, the injury with injection of small particles of dye DiI causes the phagocytic response from macrophages within 30 min after the application of the damaging effects (**Figure 2A**). This allows the use of fluorescent lipophilic carbocyanine dye DiI (1,1′-dioctadecyl-3,3,3′3′-tetramethylindocarbocyanine perchlorate, Aldrich, Sigma,

USA) as a vital nonspecific marker of microglia/macrophages. In mesencephalon (*tectum opticum*) of fish, thin needle containing crystals of dye puncture to a depth of 2–3 mm was applied. After that, the animal was placed in a separate aquarium with fresh water and with enhanced aeration for recovery. In 780 LSM microscope for multiphoton microscopy, we used lasers with pulse durations up to 10–13 (100 femtoseconds). Pulses follow with a high frequency (100 MHz), and the intervals between pulses is significantly shorter than the time ranking of the beam during scanning. The average radiation power at the same time may be small, of the same order as that of a single-photon excitation [42].

Animal at the beginning of the experiment was lying on the back, was submerged in the aquarium water, the surface of the skull was tightly pressed to the bottom wall of the special POC-R chamber. The brain of the animal was examined as a whole, without opening the skull and removing pigmented primary brain tunic. Thus, the substance for the study of DiI-labeled cells initially represented structurally heterogeneous environment, including the bones of the skull, cerebrospinal fluid, primary brain tunic, and brain tissue. The observation was carried out with special planar lens with built-in color correction (Advanced Correction System) at 20x magnification. The sample of cells was carried out in the middle portion of the optical section at a depth of 200 μm. Since the observations were made in *in vivo* mode (without production of brain sections), the scanning process have some aberrations.

After 30 min of exposure DiI to mesencephalon of juveniles *O. keta*, we observed local bright fluorescent cell bodies located in the midbrain *tegmentum* (**Figure 2A**). Cells were numerous, uniformly distributed on the depth of the investigated optical section and formed clearly a visible row of selective labeled components (**Figure 2A**).

As a result of optical scanning, we observed DiI-labeled elements without outgrowths, which formed local clusters (after 2 days) and were presented by individual elements (after 30 min) (**Figure 2B**). After 2 days in the optical sections of damaged *tegmentum*, the density of distribution of DiI-labeled cells was demonstrated as occurrence of cell conglomerates (**Figure 2B**). To investigate the space relationships of DiI-labeled cell conglomerates observed in the area of injury 2 days after injury with DiI-unlabeled, but intensely pigmented melanocytes of primary brain tunic, we spent the overlay of transmitted and fluorescent channels (**Figure 2C**). As a result of intensive multiphoton radiation, the majority of melanocytes in the primary brain tunic observed "light reaction" in which the outflow of melanin to the central part of the cell body was recorded (**Figure 2C**). Such melanocytes, devoid of outgrowths, were observed through a transmitted channel. DiI-labeled cells in deep layers of the midbrain *tegmentum* of juveniles *O. keta* were visualized through fluorescent channel and grouped into small conglomerates (**Figure 2C**). Thus, the overlay of transmitted and fluorescent channels made it possible to reconstruct three-dimensional (3D) picture of fluorescent cell conglomerates in the midbrain *tegmentum*, which was located in deep layers in combination with surface patterns of distribution of melanocytes in the primary brain tunic in the mode of *in vivo* imaging.

Based on the analysis galleries of optical sections of the midbrain of juveniles *O. keta* was created 3D reconstruction of the spatial distribution of DiI-labeled cells in the damaged area

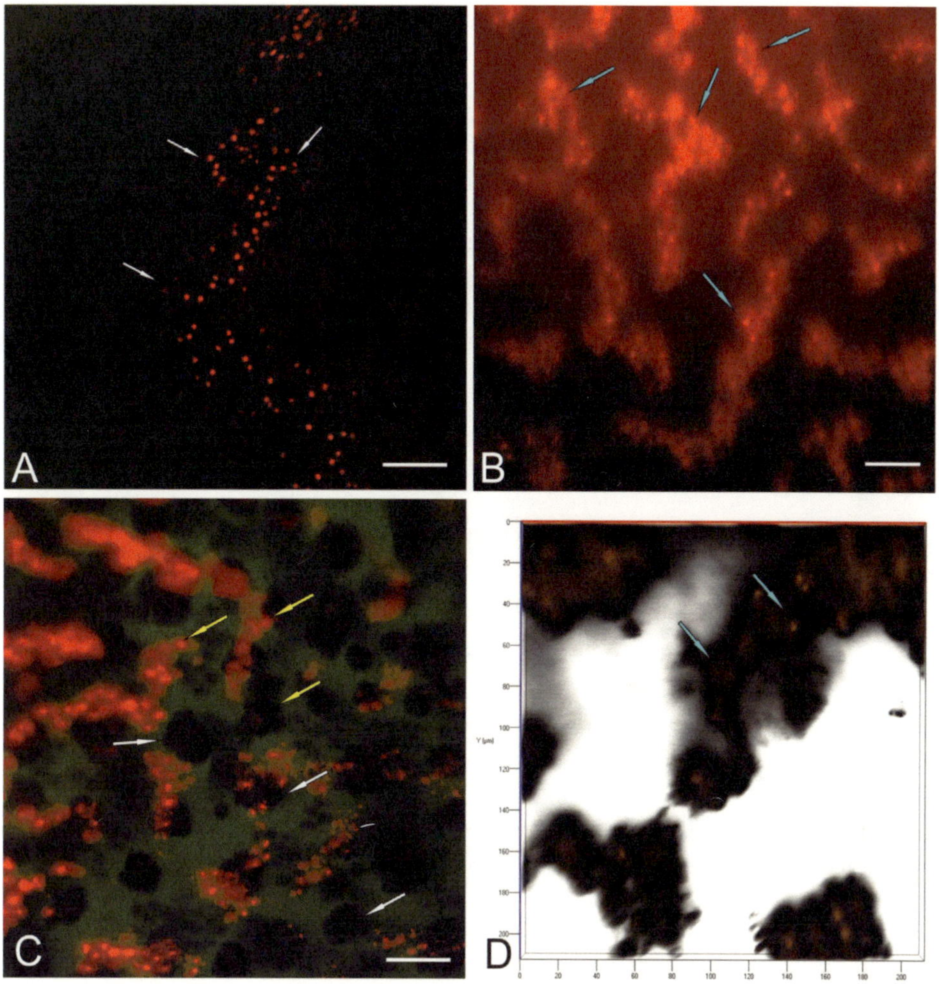

Figure 2. The results of *in vivo* monitoring at different times after injury midbrain *tegmentum* of juvenile chum salmon *Oncorhynchus keta* and injection into the area of injury carbocyanine dye (DiI). A—DiI-labeled cells (white arrows) in the surface layers of *tegmentum* 30 min after injury; B—clusters of DiI-labeled cells (arrows) 2 days after the injury; C—overly of transmitted and fluorescent channels. Transmitted channel show body of DiI-unlabeled melanocytes (white arrows) with a "light reaction." Fluorescence channel show DiI-labeled conglomerates of cells (arrows); D—3D reconstruction of 10 optical sections DiI-labeled cells in damaged *tegmentum* 2 days after injury. Multiphoton confocal microscopy. A, B, C—special planar lens (magnification 20x). Scale bar: 50 μm.

of *tegmentum* (**Figure 2D**). This 3D reconstruction shows that the distribution of fluorescent cells on day 2 after injury inflicted to *tegmentum* of juvenile *O. keta* is uneven. It revealed the formation of various DiI-labeled cell conglomerates, the number and amount of which increases toward the area of injury. Thus, for the surface layers of *tegmentum* characterized by large clusters of DiI-labeled cells (**Figure 2B**), in the deeper tegmental layers, smaller clusters of fluorescent cells were localized. Study of spatial relationships of DiI-labeled cells after 30 min of injury indicates the predominance of large accumulation of these cells in the superficial layers of the *tegmentum* and their distribution in the form of small conglomerates in the deeper layers.

In our studies, DiI was used as a dye for vital fluorescent multiphoton confocal microscopy. It can be regarded as a possible method of identifying populations of phagocytic cells in the brain, as the effective molecular markers that allow selective identification of populations of macrophages and microglia in the brain of the fish have not been developed so far. We supposed that using multiphoton confocal microscopy *in vivo* experiments allow to have the substantial preference [42]. The results of *in vivo* monitoring in different time after injury (30 min and 2 days) suggest that as a result of midbrain injury of juvenile *O. keta* has experienced rapid cellular response and the emergence of numerous stained cells in the injuries area. It is indicating active participation of such cells in migration and phagocytosis of the dye in the area of injury.

5. Cell proliferation, neural stem cells and neuronal differentiation after injury

A common feature of any regeneration-competent CNS system examined thus far is that cells lost to injury are replaced by new cells that differentiate into various cell types, including neurons. After the stab-wound lesion to the cerebellum of juvenile *O. masou*, processes of proliferation and migration of cells were amplified compared with the intact brain (**Figure 3A**). However, these processes have properties of spatial specificity, so the most proliferative activity characteristic for the dorsal matrix zone (DMZ). In this zone, proliferative activity was observed in normal (intact) conditions and we associate it with intensive persistent neurogenesis in the cerebellum of young *O. masou*. After stab-wound lesion, we verified other areas with neurogenic activity located in the dorsal part of the molecular layer, the lateral and basal regions. The emergence of neurogenic zones is attributed to the intensification of genetic programs in the proliferative neural stem cells (NSC) and the formation of local neurogenic niches (**Figure 3A, C**). Additionally to markers of neuronal differentiation and proliferation, after stab-wound lesion to the cerebellum *O. masou*, expression of doublecortin (Dcort) was detected in cells and the fibers of molecular layer. Dcort is a specific marker of migrating stem cell population, and its expression was found in neurogenic niches of molecular layer 2 days after injury. The size and location of neurogenic niches in the molecular layer of the cerebellum containing Dcort-ip cells are differed. The largest accumulation of Dcort-ip cells were found at dorsal and dorsolateral areas. Additionally to neurogenic niches, in the infraganglionic plexus of cerebellum, single Dcort-ip NSC were revealed; in thickness of the molecular layer, Dcort-ip radial glial cells were identified. So, we observed the proliferative activity in neurogenic niches combined with differentiation of some cells and their subsequent migration to the area of injury.

In the DMZ of juvenile *O. masou*, after stab-wound lesion, four types of cells labeled by PCNA have been identified. These were small, round, intensely labeled cells, or elongated ones, which are able to migrate and form the tangential and radial rows (**Figure 3B, D**). DMZ has

http://dx.doi.org/10.5772/67951

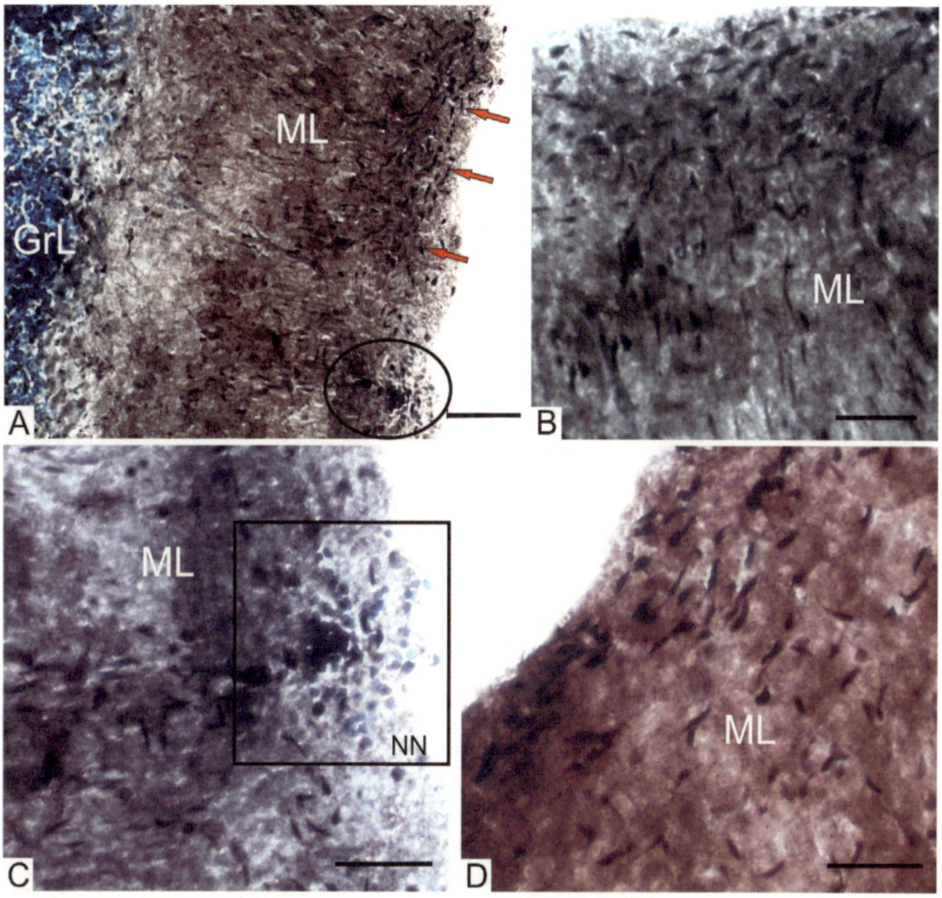

Figure 3. Localization of proliferative cell nuclear antigen (PCNA) in the cerebellum of *Oncorhynchus masou* 2 days after stab-wound injury of cerebellum. A—patterns of tangential migration (arrows) and neurogenic niche (in an oval) in the molecular layer of cerebellum; B—radial migration of PCNA-ip cells; C—neurogenic niche (in square); D—tangential migration of PCNA-ip cells. Scale bar: A—100 μm, B–D—50 μm.

been previously identified in other species of fish, in particular, *D. rerio* [28], *A. leptorhynchus* [32], *Oncorhynchus mykiss* [43]. Different species of fish do not have similar rates of proliferation in the DMZ; for example, the cerebellum of *A. leptorhynchus* contains 75% of the cells of the brain. These cells are formed mainly in the DMZ as well as granular eminentias.

After stab-wound damage to the cerebellum of juvenile *O. masou*, some neurons were eliminated by apoptosis and replaced by new cells. In the first 10 days in the damaged area, the rate of cell proliferation was increased by several times compared with other parts of the cerebellum. Experiments with BrdU labeling showed that the cells formed 2 days before the injury participate in the regeneration process [30]. This observation suggests a direct connection between the continuous cell proliferation in the intact brain and restoring of the damaged area. After injury to the fish's brain, rate of cell proliferation is much higher than in normal conditions. It is believed that some young cells develop into definitive granule neurons and subsequently most often eliminated by apoptosis. Retrograde tracing in combination with BrdU labeling S-phase of mitosis has shown that new granule neurons project to the molecular layer [30]. This fact suggests that these neurons are integrated into the existing neural network of the cerebellum.

The intensity of proliferation depends on the nature of injury and the amount of damaged brain tissue. After the damaging effects in the cerebellum of *O. masou*, proliferative activity in the cells was significantly enhanced in the dorsal area (**Figure 4A**). These data are consistent with the established data for other species of fish, in particular *A. leptorhynchus* [28] indicating that the main volume of cell proliferation after injury localizes at the DMZ. At the surface layers of the lateral zones were detected neurogenic niches (**Figure 4B**), HuCD-ip individual undifferentiated cells, and patterns of cells radial migration. Our results show that in the lateral area of cerebellum, migration processes intensified to compare with the dorsal one. Nevertheless, after a damaging effect in lateral zones were detected neurogenic niches containing HuCD-in

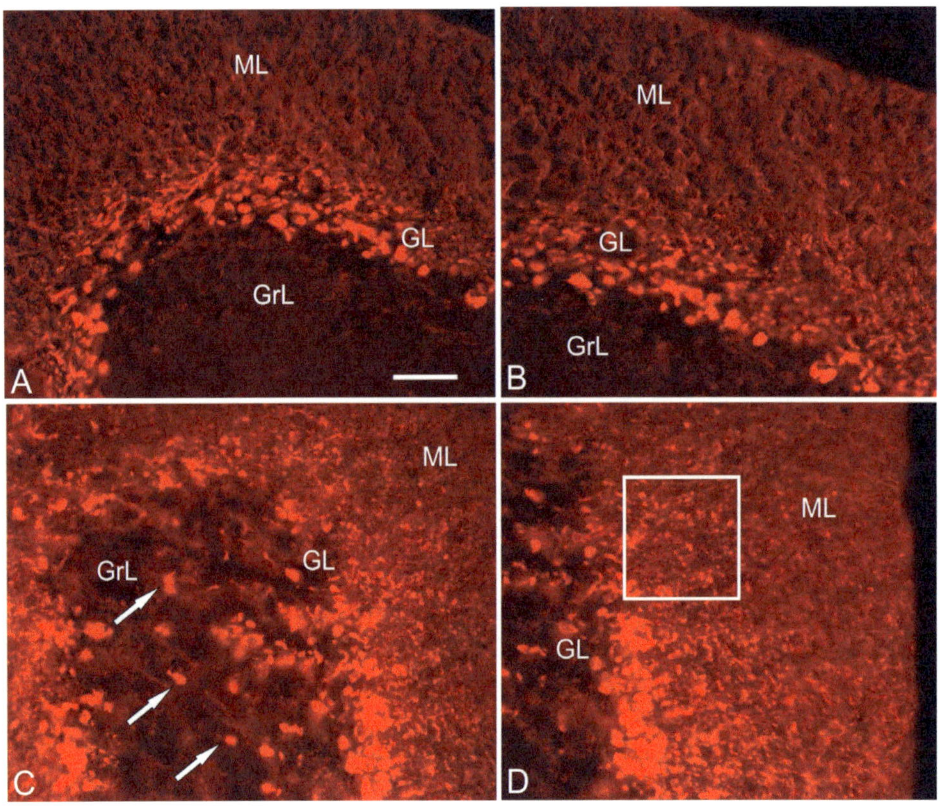

Figure 4. Immunofluorescence labeling of parvalbumin in the control (A, B) and 2 days after stab-wound injury of cerebellum (C, D) of *Oncorhynchus masou*. A—dorsal; B—lateral part of *corpus cerebellum*; Par-ip cells and fibers are present in ganglion (GL) and molecular (ML) layers, the density of Par-ip is not high; C—emergence of Par-ip cells (arrows) in the granular layer (GrL); D—increasing of density Par-ip innervation in the molecular layer (contoured by rectangle) after injury. Scale bar: 100 μm.

and Dcort-ip cells, which testifies to the existence of neural stem cells in the lateral area and enhancing their proliferative potency after injury.

In the basal area of *O. masou*, cerebellum proliferative activity of cells is high enough to control, but significantly reduced after damaging effects. The proliferative activity of the cells after the injury persists, prevailing over the processes of cell migration. A process of cell migration in basal zone is mild; however, after stab-wound lesion, the cell migration activity is increased. Thus, the proliferative activity of cells in the basal area of juvenile *O. masou* largely provides a persistent process of neurogenesis, however, after stab-wound lesion is decreased and begins to dominate the processes of cell migration.

Forebrain proliferative activity of juvenile *O. masou* revealed proliferative surface area corresponding to the periventricular area other fish, including the dorsal, lateral, and medial compartments. In addition to the large number of cell clusters in the surface zone, under conditions of normal proliferation, some cells were observed in the parenchyma of telencephalon and were identified as single or paired immunopositive PCNA-labeled cells. The presence of such cells with a high proliferative potential evidences a high level of persistent neurogenesis in telencephalon of juvenile *O. masou* salmon.

After mechanical injury in the telencephalon of *O. masou*, there has been an increase of cell density of distribution in the proliferative zone. In the deep layers of the telencephalic parenchyma of masu salmon, we observed an increase in proliferative activity: the number of single PCNA-immunopositive cells grew compared to those in intact animals. Induced neurogenesis zone appeared; it was presented by neurogenic niches and areas of secondary neurogenesis surrounded by radial glial fibers.

HuCD-immunopositive cells were identified as part external proliferative zones and in the deep layers of the telencephalon juvenile *O. masou*. We established that in juvenile *O. masou* telencephalon, HuCD protein is detected in cells being at different stages of neuronal differentiation. HuCD-immunopositive neurons were identified in the area of proliferative zone; differentiated neurons of various degrees of maturity were found in the deeper layer. The same characteristic HuCD was different levels of marking immunopositive cells. Densitometric analysis allowed to distinguish two levels of protein HuCD activity in the telencephalon *O. masou*: intense and medium. However, definitive intensely labeled neurons dominated in all areas in the control animal's brain.

We have identified four types of HuCD-labeled cells differing in morphological parameters (large and small size cell bodies) and optical density. Type 1 cells are the smallest undifferentiated cells with a high OD; Type 2 are larger oval cells with high and average value of OD; multipolar cells with high OD were the third type; and bipolar neurons with large high OD belong to the fourth type.

There are considerable changes in topography HuCD-immunopositive cells in the telencephalon *O. masou* after mechanical injury. Occurrence of neurogenic niches was registered, representing a collection of intensely labeled HuCD positive cells. Appearance of neurogenic niches was registered, representing an accumulation of intensely labeled HuCD positive cells. The density of distribution of immunopositive and negative cells increased; also, a distinct pattern of cell migration from the surface proliferative zone to the deep parenchymal layer was observed to appear.

This evidence based on HuCD and PCNA marking shows us the intense persistent neurogenesis in proliferative zone of dorsal region telencephalon *O. masou*. Distribution of neurogenic activity of deeper layers of the parenchyma appears after injury. The main sources of new neurons in the process of reparative neurogenesis are neurogenic niches.

6. Neuroprotective factors in damaged fish brain

Calbindin-D28k has been postulated to exert a neuroprotective function by buffering intracellular free Ca^{2+}. This hypothesis is supported by the findings that calbindin-D28k-expressing neurons exhibit a relative resistance to neurotoxicity induced by glutamate, calcium ionophore, or acidosis [44] and that the rate of survival of neurons can be increased after various types of insults by overexpression of the gene for calbindin-D28k [45, 46].

Other calcium-binding protein like parvalbumin may also be involved in the neuroprotective properties of fish nervous system. The results of studies on young *O. masou* showed that cerebellum after injury can significantly increase the level of expression of parvalbumin (**Figure 4A, C**). In control animals, immunofluorescence of parvalbumin was detected in cells of ganglionic layer of the cerebellum (pear-shaped Purkinje neurons) and fibers of infraganglionic plexus (**Figure 4A, B**). After mechanical injury of the cerebellum, immunofluorescence of parvalbumin was found in the cells of the granular layer and multiple synaptic terminals in the molecular layer (**Figure 4C, D**).

According to Grosche et al. [13], glutamine synthetase (GS) is a specific glial protein performing the conversion of toxic glutamate into a non-toxic amino acid, glutamine. In normal conditions, this mechanism prevents accumulation of glutamate neurotoxicity in nerve tissue, protecting neurons from cell death. But after a brain injury in a mammalian brain, volume of synthesized GS is insufficient to neutralize the toxic effects of glutamate. This determines such effects as the development of primary and secondary inflammations and progressive neurodegenerative processes observed following injury of the CNS in mammals and human glutamine synthetase [13]. In fish, the increased activity of GS is likely to provide an important mechanism for reducing the neurodegenerative effects caused by glutamate neurotoxicity. This assumption indicates the presence of certain ways that determine such strong differences in the regenerative potential of the two taxa of vertebrates [47].

The results of immunohistochemical analysis of GS indicate significant differences between the distribution of the enzyme in normal conditions and after stab-wound lesion to the cerebellum (**Figure 5A, C**). In both cases, enzyme activity was identified in cells and fibers. Densitometric analysis of enzyme activity in cells has shown that there are two levels of activity: intensive and moderate. The results of the morphological analysis and some literature data [48] indicate that cells containing the GS represent the population of astrocytes. Morphological studies of GS-ip cells in the cerebellum in young *O. masou* show the presence of a heterogeneous population of cells in control (**Figure 5A**). A maximal number of GS-ip

http://dx.doi.org/10.5772/67951

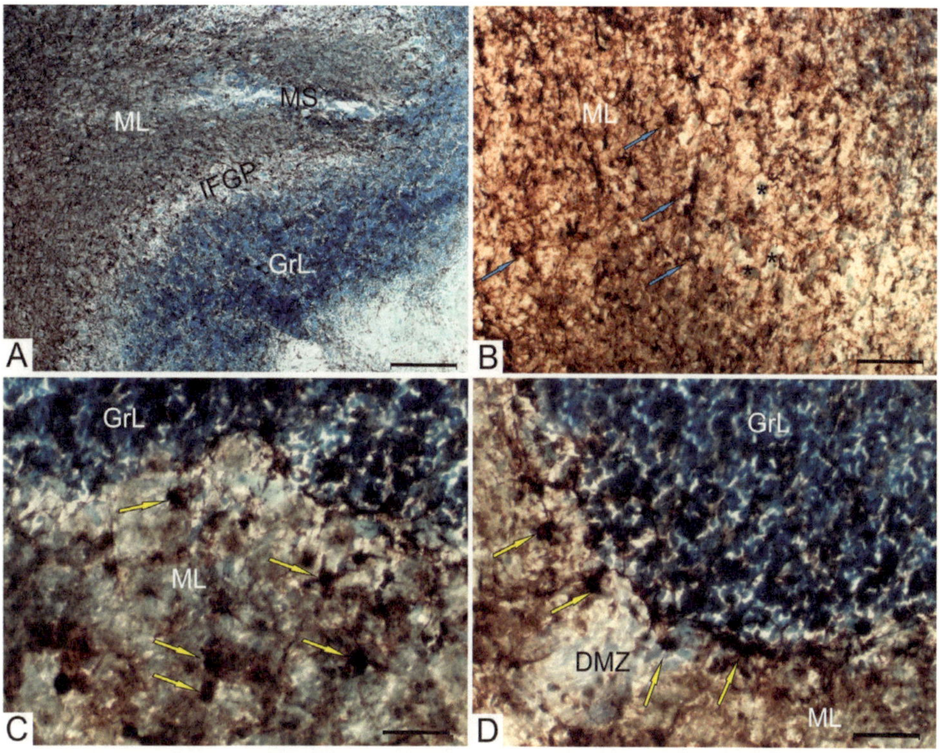

Figure 5. Immunocytochemistry of glutamine synthetase in cerebellum of *Oncorhynchus masou*. A—general view of GS-ip element distribution in the area of median suture (MS) of *corpus cerebellum*; B—GS-ip cells (arrows) in molecular layer (ML) and infraganglionic plexus (IFGP), the body of Purkinje cells indicated by asterisk. Distribution of Glutamine synthetase in the cerebellum of *Oncorhynchus masou* 2 days after stab-wound injury; C—GS-ip cells in molecular layer (arrows); D—in the dorsal matrix zone (DMZ). Scale bar: A—200 μm; B—50 μm; C, D—100 μm.

cells have been identified in the molecular layer (**Figure 5B**). The density distribution of GS-ip cells in the molecular layer in control was quite high, which indicates a high level of GS activity and, possibly, its particular impact on the processes of persistent neurogenesis.

After stab-wound lesion to the cerebellum, maximal number of GS-ip cells has been identified in the granular layer of cerebellum (**Figure 5C, D**). Redistribution of cells synthesizing GS, from the molecular to the granular layer, was revealed.

In the lateral and dorsal regions of the cerebellum of *O. masou*, increased activity of GS in the fibers was observed (**Figure 5C, D**). In the control, activity of the fibers in the granular layer (granular eminence) is not high (**Figure 5A**). We identified heterogeneous population of GS-ip cells in cerebellum of *O. masou*. We believe that among these cells are present glutamatergic neurons, containing GS labeled of metabolic glutamate and astrocytes that can receive glutamate due to its reuptake of extracellular space. The observations show that 3 days after injury in the cerebellum of *O. masou*, substantial redistribution of GS activity in various parts of the cerebellum may occur. Thus, in area of injury, we showed a significant increasing number of GS-ip cells (**Figure 5C**) and reducing number of GS-ip fibers (**Figure 5D**). This spatial specificity can be connected with both the toxicity, induced by stab-wound lesion, and change in the glutamatergic neurotransmission in damaged neural networks.

From other hand, the high activity of GS in control suggests the involvement of glutamate in plastic processes, including morphogenesis taking place during persistent neurogenesis in normal fish cerebellum [1, 47].

We believe that metabolic glutamate which is presumably involved in morphogenic cerebellar functions in *O. masou* can be localized in normal conditions, as in the growing neurons and in the fibers. This assumption is confirmed by the results of studies showing the high activity of GS in the cells of the molecular layer, especially in dorsal region. This region contains a DMZ of cerebellum, characterized by a high neurogenic activity (**Figure 5D**). Our results suggest that the pattern of GS activity was decreased during 3 days of injury, but we do not exclude the possibility that these changes are temporary. This hypothesis was supported by data of enzyme immunoassay (ELISA) carried out by us in the cerebellum of *O. masou* (**Figure 6**). Thus, during long-term monitoring, the GS activity was found to be increased during the initial few hours after injury (1–3 h) and then continue to increase till the 14th day, except a decrease in the 12 h after injury. However, these changes of metabolic activity of GS may only represent a local decrease in enzyme activity as shown on the third day. The results of ELISA immunoassay established that the enzyme activity after damaging effects has a complicated pattern. The increase in enzyme activity was observed during the first few hours (1–3 h) after injury and at second, fifth, and tenth days after the damaging effects (**Figure 6**). On the second day, we observed increase in activity of GS by ELISA immunoassay, which is consistent with results of IHC labeling on frozen brain sections. Thus, the decrease in activity of GS on the third day after mechanical injury of cerebellum *O. masou* can be a particular manifestation of the changes in the metabolic status. At sufficiently high intensity of persistent neurogenesis in young *O. masou*, we tend to believe that the response from the GS-producing elements

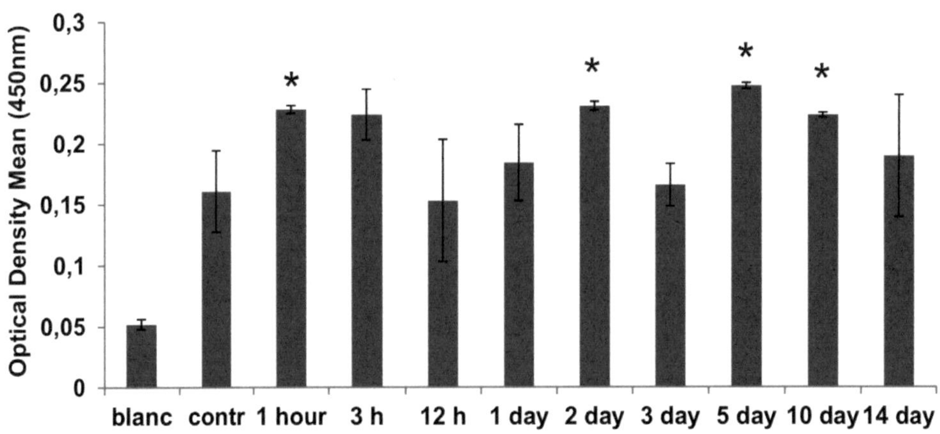

Figure 6. The immunoassay data of glutamine synthetase content in the cerebellum of *Oncorhynchus masou* at different time intervals after stab-wound injury. Protein concentration is 100 ng/ml. On the *x* axis are shown different time points after injury, whereas mean optical density is shown on *y* axis—(absorbance at 450 nm). Data are shown as mean ± S.E.M; *P < 0.05 significant differences compared to the control group (n = 5 in each group).

in the cerebellum can be complex and ambiguous. So, damaging effects can be the cause of death of a large number of cells by apoptosis, as a result of glutamate toxicity. However, the presence of increased level of GS in first, second, and fifth day indicates a rather high-level production of the enzyme whose activity can be reduced and cyclically be determined by different factors, the nature of which remains to be established.

The increase in the number of GS-ip cells in the granular layer and the high activity of GS in them on the second and third day are referred by us to as the astrocytic response observed after a damaging impact. However, the number of these cells is not high enough for response to the GS-ip cells as «reactive gliosis» appeared on central nervous system of mammals after damaging effects. In the mammalian brain, as is known, as a result of a traumatic impact, pool of reactive astrocytes formed morphological and biochemical barrier, features which significantly differ from those in normal astrocytes [47]. The cellular mechanism associated with the transformation of a population of astrocytes and isolating a subpopulation of activated glia in the brain of the fish is currently poorly understood. Unlike mammalian brain, astrocytes in a fish brain do not form the astrocytic barrier, which is characteristic for the development of posttraumatic process in the mammalian brain. Nevertheless, changing GS synthesis is an unambiguous evidence in favor of the neuroprotective properties of the enzyme and increased production in the cerebellum *O. masou*. This indicates that not only GS is the marker of cells involved in the conversion of glutamine/glutamate but it can also be considered as an effective neuroprotective factor contributing to posttraumatic reparative processes.

7. Conclusion

The fish brain has a unique feature of vertebrates—it grows with the growth of body over a lifetime. In this regard, fish is a convenient model for the study of embryonic and postembryonic development of the central nervous system and of the influence of different factors on these processes. Injury of fish brain creates special conditions for the implementation of genetic programs aimed at strengthening the proliferation of progenitor cells as well as activation and proliferation activity in the neuronal stem cells. Study of neurogenic activity, migration, and differentiation in the neurogenic niches contributes to a better understanding about how these structures operate, not only in fish but in other vertebrates as well.

Acknowledgements

This study was supported by a grant of the President of the Russian Federation (MD-4318.2015.4), Grant of Scientific Foundation (№ 10046–2016.4), and the "Far East" Program for Basic Research of the Far East Branch of the Russian Academy of Sciences 2015–2017 (project number 15-I-6-116, Section III).

We are grateful to Dr. Sachin Shukla Eye Research Centre (Hyderabad, India) for the participation in ELISA immunoassay and editing the text.

Author details

Evgeniya V. Pushchina[1,2]*, Anatoly A. Varaksin[1], Mariya E. Stukaneva[1] and Eva I. Zharikova[1]

*Address all correspondence to: puschina@mail.ru

1 National Scientific Center of Marine Biology, Far Eastern Branch, Russian Academy of Science, Laboratory of Cell Differentiation, Vladivostok, Russia

2 Bogomoletz Institute of Physiology, National Academy of Science of Ukraine, Laboratory of Stem Cell Biology, Kiev, Ukraine

References

[1] Zupanc GK, Sîrbulescu RF. Teleost fish as a model system to study successful regeneration of the central nervous system. Cur Top Microbiol Immunol 2013; 367: 193–233.

[2] Zupanc GK. Towards brain repair: insights from teleost fish. Semin Cell Dev Biol 2009; 20(6): 683–690.

[3] Pushchina EV, Varaksin AA, Obukhov DK. Participation of neurochemical signaling in adult neurogenesis and differentiation. In: Heinbockel T, editor. Neurochemistry. Rijeka: Intech; 2014, pp. 225–255.

[4] Zupanc GK, Horschke I. Proliferation zones in the brain of adult gymnotiform fish: a quantitative mapping study. J Comp Neurol 1995; 353(2): 213–233.

[5] Zikopoulos B, Kentouri M, Dermon CR. Proliferation zones in the adult brain of a sequential hermaphrodite teleost species (*Sparus aurata*). Brain Behav Evol 2000; 56(6): 310–322.

[6] Extröm P, Johnsson CM, Ohlin LM. Ventricular proliferation zones in the brain of an adult teleost fish and their relation to neuromeres and migration (secondary matrix) zones. J Comp Neurol 2001; 436(1): 92–110.

[7] Zupanc GK, Hinsch K, Gage FH. Proliferation, migration, neuronal differentiation, and long-term survival of new cells in the adult zebrafish brain. J Comp Neurol 2005; 488(3): 290–319.

[8] Grandel H, Kaslin J, Ganz J, Wenzel I, Brand M. Neural stem cells and neurogenesis in the adult zebrafish brain: origin, proliferation dynamics, migration and cell fate. Dev Biol 2006; 295(1): 263–277.

[9] Fernández AS, Rosillo JC, Casanova G, Olivera-Bravo S. Proliferation zones in the brain of adult fish *Austrolebias* (Cyprinodontiform: Rivulidae): a comparative study. Neuroscience 2011; 189: 12–24.

[10] Zupanc GK, Kompass KS, Horschke I, Ott R, Schwarz H. Apoptosis after injuries in the cerebellum of adult teleost fish. Exp Neurol 1998; 152(2): 221–230.

[11] Becker CG, Becker T. Adult zebrafish as a model for successful central nervous system regeneration. Restor Neurol Neurosci 2008; 26(2): 71–80.

[12] Palmer AM, Marion DW, Botscheller ML, Bowen DM, DeKosky ST. Increased transmitter amino acid concentration in human ventricular CSF after brain trauma. Neuroreport 1994; 6(1): 153–156.

[13] Grosche J, Härtig W, Reichenbach A. Expression of glial fibrillary acidic protein (GFAP), glutamine synthetase (GS), and Bcl-2 protooncogene protein by Müller (glial) cells in retinal light damage of rats. Neurosci Lett 1995; 185(2): 119–122.

[14] Kerr JF, Gobé GC, Winterford CM, Harmon BV. Anatomical methods in cell death. In: Schwartz LM, Osborne BA, editors. Cell death. San Diego: Academic Press; 1995, pp. 1–27.

[15] Vajda FJ. Neuroprotection and neurodegenerative disease. J Clin Neurosci 2002; 9(1): 4–8.

[16] Hitchcock P, Ochocinska M, Sieh A, Otteson D. Persistent and injury-induced neurogenesis in the vertebrate retina. Prog Retin Eye Res 2004; 23(2): 183–194.

[17] Becker CG, Becker T. Growth and path finding of regenerating axons in the optic projection of adult fish. J Neurosci Res 2007; 85(12): 2793–2799.

[18] Zupanc GK, Sîrbulescu RF. Adult neurogenesis and neuronal regeneration in the central nervous system of teleost fish. Eur J Neurosci 2011; 34(6): 917–929.

[19] Liou AK, Clark RS, Henshall DC, Yin X-M, Chen J. To die or not to die for neurons in ischemia, traumatic brain injury and epilepsy: a review on the stress-activated signaling pathways and apoptotic pathways. Prog Neurobiol 2003; 69(2): 103–142.

[20] Elmore S. Apoptosis: a review of programmed cell death. Toxicol Pathol 2007; 35(4): 495–516.

[21] Nagamoto-Combs K, McNeal DW, Morecraft RJ, Combs CK. Prolonged microgliosis in the rhesus monkey central nervous system after traumatic brain injury. J Neurotrauma 2007; 24(11): 1719–1742.

[22] Weber JT. Altered calcium signaling following traumatic brain injury. Front Pharmacol 2012; 3: 60. http://dx.doi.org/10.3389/fphar

[23] Zupanc MM, Zupanc GK. Upregulation of calbindin-D28k expression during regeneration in the adult fish cerebellum. Brain Res 2006; 1095(1): 26–34.

[24] Zupanc MM, Wellbrock UM, Zupanc GK. Proteome analysis identifies novel protein candidates involved in regeneration of the cerebellum of teleost fish. Proteomics 2006; 6: 677–696.

[25] Lewis GP, Guérin CJ, Anderson DH, Matsumoto B, Fisher SK. Rapid changes in the expression of glial cell. Proteins caused by experimental retinal detachment. Am J Ophthalmol 1994; 118(3): 368–376.

[26] Stukaneva ME, Pushchina EV. Proliferation and apoptosis in the cerebellum of Pacific salmon fry after mechanical damage. Neurophysiology 2015; 47(5–6): 389–399.

[27] Zupanc GK. Neurogenesis, cell death and regeneration in the adult gymnotiform brain. J Exp Biol 1999; 202(Pt 10): 1435–1446.

[28] Zupanc GK. Neurogenesis and neuronal regeneration in the adult fish brain. J Comp Physiol A Neuroethol Sens Neural Behav Physiol 2006; 192(6): 649–670.

[29] Candal E, Anadón R, Bourrat F, Rodríguez-Moldes I. Cell proliferation in the developing and adult hindbrain and midbrain of trout and medaka (teleosts): a segmental approach. Brain Res Dev Brain Res 2005; 160(2): 157–175.

[30] Zupanc GK, Ott R. Cell proliferation after lesions in the cerebellum of adult teleost fish: time course, origin, and type of new cells produced. Exp Neurol 1999; 160(1): 78–87.

[31] Puschina EV, Obukhov DK, Varaksin AA. Features of adult neurogenesis and neurochemical signaling in the Cherry salmon *Oncorhynchus masou* brain. Neural Regen Res 2013; 8(1): 13–23.

[32] Hinsch K, Zupanc GK. Generation and long-term persistence of new neurons in the adult zebrafish brain: a quantitative analysis. Neuroscience 2007; 146(2): 679–696.

[33] Pouwels E. On the development of the cerebellum of the trout, *Salmo gairdneri*: I. Patterns of cell migration. Anat Embryol (Berl) 1978; 152(3): 291–308.

[34] Stukaneva ME, Pushchina EV, Varaksin AA, Shukla S. Monitoring of cell migration and apoptosis in cerebellum of juvenile masu salmon *Oncorhynchus masou* after injury. Am J BioSci 2015; 3(2–3): 19–27.

[35] Puschina EV, Obukhov DK. Processes of proliferation and apoptosis in the brain of the Amur sturgeon. Neurophysiology 2011; 43(4): 271–286.

[36] Rajendran RS, Zupanc MM, Lösche A, Westra J, Chun J, Zupanc GK. Numerical chromosome variation and mitotic segregation defect in the adult brain of teleost fish. Dev Neurobiol 2007; 67(10): 1334–1347.

[37] Bengzon J, Kokaia Z, Elmér E, Nanobashvili A, Kokaia M, Lindvall O. Apoptosis and proliferation of dentate gyrus neurons after single and intermittent limbic seizures. Proc Natl Acad Sci USA 1997; 94(19): 10432–10437.

[38] Zupanc GK, Clint SC, Takimoto N, Hughes AT, Wellbrock UM, Meissner D. Spatiotemporal distribution of microglia/macrophages during regeneration in the cerebellum of adult teleost fish, *Apteronotus leptorhynchus*: a quantitative analysis. Brain Behav Evol 2003; 62(1): 31–42.

[39] Ayari B, El Hachimi KH, Yanicostas C, Landoulsi A, Soussi-Yanicostas N. Prokineticin 2 expression is associated with neural repair of injured adult zebrafish telencephalon. J Neurotrauma 2010; 27(5): 959–972.

[40] Kroehne V, Freudenreich D, Hans S, Kaslin J, Brand M. Regeneration of the adult zebrafish brain from neurogenic radial glia-type progenitors. Development 2011; 138(22): 4831–4841.

[41] Craig SE, Calinescu AA, Hitchcock PF. Identification of the molecular signatures integral to regenerating photoreceptors in the retina of the zebrafish. J Ocul Biol Dis Infor 2008; 1(2–4): 73–84.

[42] Pushchina EV, Varaksin AA, Shukla S, Bulygyn DA. Multiphoton confocal microscopy (*in vivo* imaging) in the study of early response of macrophages/microglia in damaged midbrain of juvenile chum salmon *Oncorhynchus keta*. Am J BioSci 2015; 3(2–3): 12–18.

[43] Pushchina EV, Varaksin AA, Obukhov DK. Reparative neurogenesis in the brain and changes in the optic nerve of adult trout *Oncorhynchus mykiss* after mechanical damage of the eye. Rus J Dev Biol 2016; 47(1): 11–32.

[44] Mattson MP, Rychlik B, Chu C, Christakos S. Evidence for calcium-reducing and excitoprotective roles for the calcium-binding protein calbindin-D28k in cultured hippocampal neurons. Neuron 1991; 6(1): 41–51.

[45] D'Orlando C, Celio MR, Schwaller B. Calretinin and calbindin D-28 k, but not parvalbumin protect against glutamate-induced delayed excitotoxicity in transfected N18-RE 105 neuroblastoma-retina hybrid cells. Brain Res 2002; 945(2): 181–190.

[46] Phillips RG, Meier TJ, Giuli LC, McLaughlin JR, Ho DY, Sapolsky RM. Calbindin D28K gene transfer via herpes simplex virus amplicon vector decreases hippocampal damage *in vivo* following neurotoxic insults. J Neurochem 1999; 73(3): 1200–1205.

[47] Zupanc GK. Adult neurogenesis and neuronal regeneration in the teleost fish brain: implications for the evolution of a primitive vertebrate trait. In: Bullock TH, Rubenstein LR, Zupanc GK, Sîrbulescu RF, editors. The evolution of nervous systems in non-mammalian vertebrates. Oxford: Academic Press; 2006, pp. 485–520.

[48] Revett TJ. Glutamate system, amyloid ß peptides and tau protein: functional interrelationships and relevance to Alzheimer disease pathology. J Psychiatry Neurosci 2013; 38(1): 6–23.

Chapter 8

Effect of Local Delivery of GDNF Conjugated Iron Oxide Nanoparticles on Nerve Regeneration along Long Chitosan Nerve Guide

Federica Fregnan, Michela Morano, Ofra Ziv-Polat, Mira M. Mandelbaum-Livnat, Moshe Nissan, Tolmasov Michael, Akiva Koren, Tali Biran, Yifat Bitan, Evgeniy Reider, Mara Almog, Nicoletta Viano, Shimon Rochkind, Stefano Geuna and Abraham Shahar

Additional information is available at the end of the chapter

http://dx.doi.org/10.5772/intechopen.68526

Abstract

Local delivery of neurotrophic factors is a pillar of neural repair strategies in the peripheral nervous system. The main disadvantage of the free growth factors is their short half-life of few minutes. In previous studies, it was demonstrated that conjugation of various neurotrophic factors to iron oxide nanoparticles (IONP) led to stabilization of the growth factors and to the extension of their biological activity compared to the free factors. In vitro studies performed on organotypic dorsal root ganglion (DRG) cultures seeded in NVR gel (composed mainly of hyaluronic acid and laminin) revealed that the glial cell–derived neurotrophic factor (GDNF) conjugated to IONP-enhanced early nerve fiber sprouting and accelerated the onset and progression of myelin significantly earlier than the free GDNF and other free and conjugated factors. The present article summarizes results of in vivo study, aimed to test the effect of free versus conjugated GDNF on regeneration of the rat sciatic nerve after a severe segment loss. We confirmed that nerve device enriched with a matrix with GDNF gives more successful results in term of regeneration and functional recovery in respect to the hollow tube; moreover, there are no detectable differences between free versus conjugated GDNF.

Keywords: GDNF conjugated iron oxide nanoparticles, chitosan tube, peripheral nerve regeneration

1. Introduction

The incidence of nerve injury is quite high in the world, related also to the exposition of nerve tissue in the body. Despite the good regeneration ability retained by peripheral nerve system in adulthood, nerve injuries are associated with high morbidity and deep alteration of patient's life [1]. Severe transection injuries, in which nerve continuity is lost, require a surgical approach, specially where the distance between the proximal and the distal nerve stumps is extended. In order to avoid negative side effect of the autologous nerve graft, the current gold standard technique, and to improve the functional recovery and reduce neuropathic pain, various biosynthetic nerve grafts have been developed to bridge the two nerve stumps [2–4]. Some of the artificial nerve guides have been approved for clinical use, and indeed, their use in reconstructive surgery is restricted because they still show limitations, principally for long-distance repair [5]. One category of these nerve conduits is defined absorbable conduit, according to their characteristic to be degraded in the host body; among these, the chitosan tube has been widely used in pre-clinical studies obtaining promising results [4, 6, 7]. Chitosan is a polysaccharide derived from chitin, it is biocompatible, and it can promote glial cells survival and neurite outgrowth [8–10]. It has been shown that hollow chitosan tube can be effective as autologous nerve graft for bringing 10-mm gap in rat sciatic nerve model [7, 11, 12]. In order to increase the performance of chitosan tube for long defect repair; here, we filled the tube with a hydrogel, called NVR gel, composed mainly by hyaluronic acid and laminin. Previous works demonstrated that NVR gel is a promising hydrogel that allow neurite outgrowth and cell survival in vitro [13]. Furthermore, we used NVR gel as a carrier for neurotrophic factors. In particular, we focus our attention on glial cell line–derived neurotrophic factor (GDNF), which is mainly involved in motor neuron regrowth and remyelination [14–16]. After nerve injury, endogenous growth factors are released by neuronal and glial cells from the distal nerve stump, and they can stimulate and guide axon regeneration; yet, this support is ineffective in regeneration over long distance and extended in time due to the short life of neurotrophic factors and the decline of their production [17, 18]. The implementation of the nerve conduit with neurotrophic factors has the aim to maintain elevated the concentration of these growth factors and to extend their action for the prolonged time needed for axons to reach the target. For this purpose, we use the NVR gel as a scaffold for the release of GDNF conjugated to iron oxide nanoparticles (GDNF-IONP). We have previously demonstrated that conjugation of various neurotrophic factors to iron oxide nanoparticles (IONP) led to stabilization of the growth factors and to prolonged biological activity respect to the non-conjugated factors [19]. IONP are considered biocompatible and biodegradable, they are actually used for various biomedical application [20–22], and thus, they were suggested for in vivo use for peripheral nerve regeneration [23]. Moreover, in vitro experiments showed that the combination of NVR gel and IONP represents a permissive environment to neurite outgrowth [19], and GDNF-IONP has been shown to accelerate the onset and progression of myelin in organotypic dorsal root ganglion (DRG) cultures seeded in NVR gel, significantly earlier than the free GDNF and other free and conjugated factors [24].

In the present study, we explored the efficacy of chitosan tube enriched with NVR gel and GDNF-IONP to the repair of critical length, 15 mm, sciatic nerve defect in rat model. We evaluated the functional and morphological outcome of nerve regeneration at 5 months after nerve injury, analyzing the effects given by the presence of NVR gel alone, NVR gel with GDNF-IONP or free GDNF inside the chitosan tube respect to the empty nerve device.

2. Materials and methods

2.1. Synthesis of GDNF-IONP

Dextran-coated iron oxide magnetic nanoparticles of 10 nm diameter were prepared as described previously [3]. The dextran coating was used to covalently bind the protein GDNF using divinyl sulfone binding reagent [1].

2.2. NVR gel preparation

NVR gel is a matrix supporting cell growth and survival composed by high-molecular-weight hyaluronic acid (3×10^6 Da, BTG Polymers, Kiryat Malachi, Israel) and laminin (Sigma, Rehovot, Israel) [4]. For the in vivo application as a tube filler, the NVR gel was diluted in nutrient medium corresponding to Dulbecco's modified eagle medium-nutrient mixture F-12 (DMEM-F12), supplemented with 10% fetal calf serum (FCS), 2-nM glutamine, 6-g/L D-glucose, 25-μg/mL gentamycine and 50-ng/mL IGF-I. The final concentration of the solution is 0.5% to render more suitable gel manipulation. NVR gel was filled into the chitosan tubes during the surgical implantation using a syringe.

2.3. Preparation of chitosan conduit

Chitosan tubes were manufactured by Medovent GmbH (Mainz, Germany) under ISO 13485 conditions from chitin tubes made following three main procedures: the extrusion process, distinctive washing, and hydrolysis steps to reach the required medium degree of acetylation (DAII). Tubes were finally cut into the length of 17 mm and treated with ethylene oxide for sterilization.

2.4. Animals

All animal experiments were approved by the Institutional Animal Care and Usage Committee (IACUC) and adhered strictly to the Animal Care guidelines. Female Wistar rats were brought to the vivarium 2 weeks prior to the surgery and housed two per cage with a 12-h light/dark cycle, with free access to food and water.

2.5. Experiment design and surgical technique

Fifty female Wistar rats, weighing 200–250 g each, were studied using an experimental model for producing a complete peripheral nerve injury with massive nerve defect that has recently been described [7]. During the 2 weeks before surgery and for the entire study, the rats were given Amitriptyline in their drinking water, in order to decrease autotomy—self-eating of toes—in the operated limb. Before surgery, a general anesthesia was induced with intraperitoneal injection of xylazine (15 mg/kg) and ketamine (50 mg/kg).

During the surgical procedures, a high magnification microscope was used. The left sciatic nerve was uncovered and disconnected from biceps femoris and semimembranous muscles, beginning from the area of branches to the glutei and hamstring muscles and distally to the trifurcation into peroneal, tibial, and sural nerves. At the third femur level, the sciatic nerve was fully transected using microsurgical scissors, and a 15-mm nerve segment was removed.

Afterward, the rats were divided into five experimental groups according to the type of implant: empty tubes ($n = 10$); tubes filled with NVR gel ($n = 10$); tubes filled with NVR gel enriched with free GDNF ($n = 10$); tubes filled with NVR gel enriched with GDNF-IONP ($n = 10$); and autologous nerve grafts ($n = 10$).

For nerve reconstruction in treatment groups, a 17-mm chitosan empty tube was located between the proximal and the distal sides of the transected sciatic nerve. Both proximal and distal sides of the sciatic nerve were positioned 1 mm into the tube ends, providing a 15-mm nerve gap between the proximal and distal ends. Then, the tube ends were sutured to the epineurium at the proximal and distal nerve stumps using a 9-0 nonabsorbable suture.

In control group (autologous nerve graft reconstruction), after exposition the left sciatic nerve, a 15-mm nerve segment was severely cut, using micro scissors, at the femur level, below the superior gluteal nerve and above the dissection of the sciatic nerve into the tibial and peroneal nerves. Immediately thereafter, inverse end-to-end coaptation of the nerve segment was performed using 2–3 nonabsorbable 10-0 sutures. Coaptation of nerve fascicles was performed to preserve all the fascicles within the epineural sac. Then, the muscular, subcutaneous, and skin layers were closed.

2.6. End-point assessments

Assessments before and after surgery (30, 90, and 150 days postoperatively) were carried out in a blinded manner without disclosure of rat's affiliation to the evaluating team. The assessments consisted of functional motor assessment of the sciatic nerve utilizing sciatic function index (SFI), somatosensory evoked potentials (SSEP), ultrasound evaluation, and morphological analysis.

2.6.1. Electrophysiological evaluation

SSEPs were recorded in both the operated and intact limbs using a Dantec™ KEYPONT® PORTABLE. Conductivity of the sciatic nerve was measured by stimulating the sciatic nerve at the level of the tarsal joint with simultaneous recording from the skull over the somatosensory cortex in anesthetized rats. Two subcutaneous needle electrodes were placed under the skin of the skull, when the active electrode was placed above the somatosensory cortex along the midline, and the reference electrode was placed between the eyes. The ground electrode was placed subcutaneously on the dorsal neck. Stimulation of the sciatic nerve was conducted by a set of two polarized electrodes located on the lateral aspect of the tarsal joint. The sciatic nerve was stimulated by 300 pulses of 0.2 ms in duration with a rate of 3 Hz. The intensity of the stimulus was set to 2–5 mA, causing a slight twitching of the limb under observation. A response to the stimulus was considered positive if an evoked potential appeared in at least two consecutive tests.

2.6.2. Ultrasound evaluation (US)

During the observational period, imaging studies employing ultrasonography were carried out, in anesthetized rats, for real-time evaluation of nerve regeneration inside the chitosan tubes. The lateral aspect of the right leg was shaved to improve transducer contact. The sonographic scanning technique included longitudinal and transverse sections. US examinations were performed using conventional US units equipped with color Doppler capabilities using 7–15 MHz linear transducer, yielding an axial resolution of 0.2–0.4 mm. The identification of the chitosan conduit on the US image was based on the recognition of a hyperechoic structure of a tubular shape in the longitudinal axis and a circular shape on the transverse section.

2.6.3. Sample resin embedding

All the nerve samples were harvested 5 months after the surgical implantation. The complete tube was collected taking care to preserve part of the proximal and the distal nerve and was fixed for 4–6 h at 4°C in 0.1-M phosphate buffer (pH 7.4) with 2.5% glutaraldehyde. After the fixation, all the tubes were opened using a scalpel and removed in order to free and clean the regenerated nerve inside the tube and prepared it to be processed for resin inclusion. The post-fixation of nerves regenerated inside tubes was done using 2% osmium tetroxide for 2 h, and then, the samples were dehydrated in ethanol from 30 to 100%. Two washings of 7 m using propylene oxide were performed, and then, samples were embedded in a mixture of propylene oxide and Glauerts' mixture of resins (50% Araldite HY964, 50% Araldite M and 0.5% dibutylphthalate) mixed in equal parts and left 1 h at room temperature. A second embedding with Glauerts' mixture of resins alone for an overnight was performed. The Glauerts' mixture of resins was changed, and samples were left 37°C for 1 h. Two following samples embedding were done using resin with 2% of accelerator 964. Samples were left for 3 days at 60°C.

For stereological analysis, resin-embedded nerve was cut from the distal stump using Ultracut UCT ultramicrotome (Leica Microsystems, Wetzlar, Germany) to obtain transverse semi-thin sections (2.5 μm of thickness) for optical analysis.

2.6.4. Morphometrical analysis

Semi-thin transverse nerve sections were stained with 1% toluidine blue and analyzed in high-resolution light microscopy. The qualitative and quantitative morphological analysis was performed with DM4000B microscope and DFC320 digital camera, using IM50 image manager system (Leica Microsystems). One randomly selected semi-thin section was used to measure the total cross-section area of the nerve. Using a systematic protocol described in [25], 12–16 fields were selected, and the following parameters were measured: mean fiber density (number of fiber/field area), total number of myelinated fibers (mean fiber density ×

area of the nerve section) fiber area, axon area, fiber diameter (D) and fiber axon (d), myelin thickness [$(D - d)/2$], and g-ratio (d/D).

2.7. Statistical analysis

Functional and electrophysiological analysis and calculations were done using MatLab software (Ver. 2008b, The MathWorks, Inc.). Nonparametric statistics were used in this study. Hence, all figures are presented with median ± mad. Significance levels were calculated using a Mann-Whitney U test and a Wilcoxon signed rank test. SSEP responses were analyzed as categorical parameters using χ^2 test.

For the stereological analysis, more than five sciatic nerves were analyzed for each single groups (autograft, $n = 5$; tube with NVR gel, $n = 8$; tube with NVR gel enriched with GDNF-IONP, $n = 8$; tube with NVR gel enriched with free GDNF, $n = 7$). The ANOVA one way followed by Bonferroni post hoc test was performed using SPSS Statistic Program. Results are reported as mean +SD.

3. Results

All the five defined animal groups (autograft, DAII empty tube, DAII tube with NVR gel, DAII tube with NVR gel and GDNF-IONP, DAII tube with NVR gel with GDNF) were followed-up until 5 months after nerve injury. During the follow-up period, the regeneration and the position of the implanted tubes were evaluated using ultrasound observation that enabled to observe in vivo the condition of the tube without scarifying the rats. All implants were found to be complete and in correct position (**Figure 1**).

Five months after nerve guide implantation, animals were sacrificed, and all the chitosan tubes were removed for nerve regeneration evaluation. For morphological and morphometrical analysis, the autograft group was considered as the control group, since in the comparison with healthy nerve already revealed the best regenerative aspect [26, 27].

A nerve regeneration was observed in all of the investigated conditions: a higher percentage of regenerated nerves was detected for tubes containing NVR gel with GDNF-IONP (100%), while a lower percentage of regenerated nerves (56%) was observed for tubes containing NVR

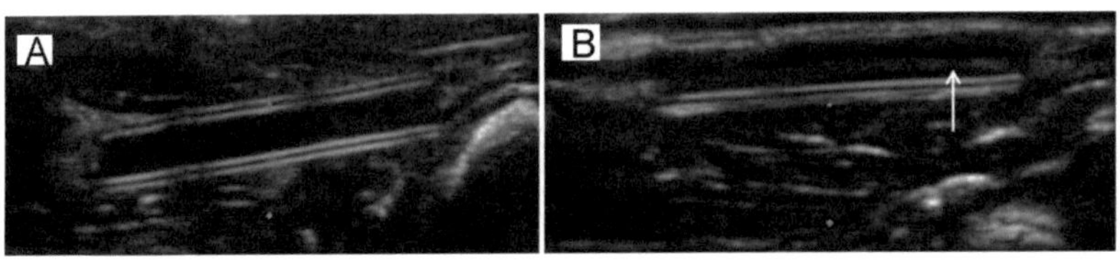

Figure 1. Ultrasound evaluation of implanted chitosan tube. (A) 20-day post-tube implantation—no regeneration is visible inside the tube and (B) 90 days post-tube implantation—nerve filaments are present inside the tube reconnecting nerve stumps (white arrows).

gel with GDNF; the lowest percentage of regenerated nerves was found for empty tubes (11%) or tubes filled with NVR gel alone (30%).

3.1. Functional analysis

In order to assess the functional recovery, we used the electrophysiological evaluation. We recorded SSEP at different time points (0, 30, 90 and 150 days after injury) calculating SSEP peak to peak (P2P) amplitude (NR type) in which the operated limb was compared to the intact limb (**Figure 2**). Regarding the operated limb, all groups showed a decrease at the first follow-up (30 days), followed by an increase, except for the tube filled with NVR gel and NVR gel enriched with GDNF-IONP, which recovered only after 90 days. Interestingly, for what concerns the intact limb, the animal group treated with tube filled with NVR gel enriched with GDNF-IONP also exhibited a decrease in amplitude following the operation and then recovered, while both the autologous nerve graft reconstruction and the tube filled with NVR gel enriched with free GDNF groups demonstrated an increase in P2P.

It was interesting to observe the comparison of the P2P among the different treatments in both the operated and intact limbs (**Figure 3**). At 30 days after injury, P2P amplitude is significantly higher in the group of animals in which truncated nerves were repaired with tubes filled with NVR gel and GDNF-IONP respect to those repaired with autologous nerve graft ($p < 0.05$).

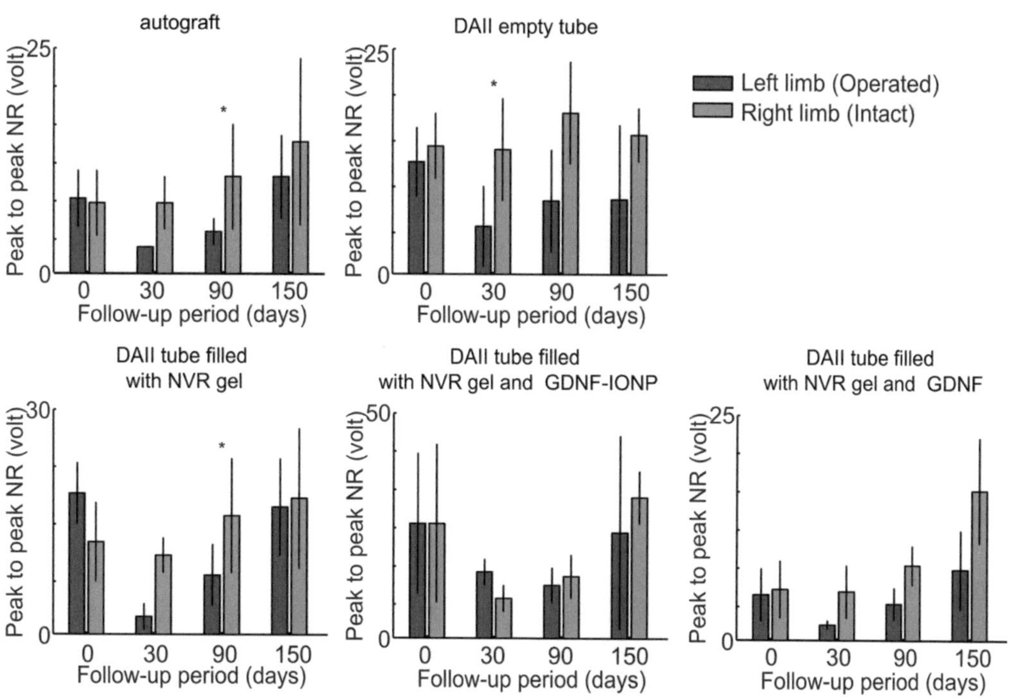

Figure 2. Comparison of somatosensory evoked potentials (SSEP) peak-to-peak (P2P) amplitude among various treatments. Data were gathered from each limb separately (operated and intact) at four follow-up periods: 0 (pre-operatively), 30, 90, and 150 days. Values are presented as median ± mad. Statistical analysis: Wilcoxon signed rank test. *$p < 0.05$.

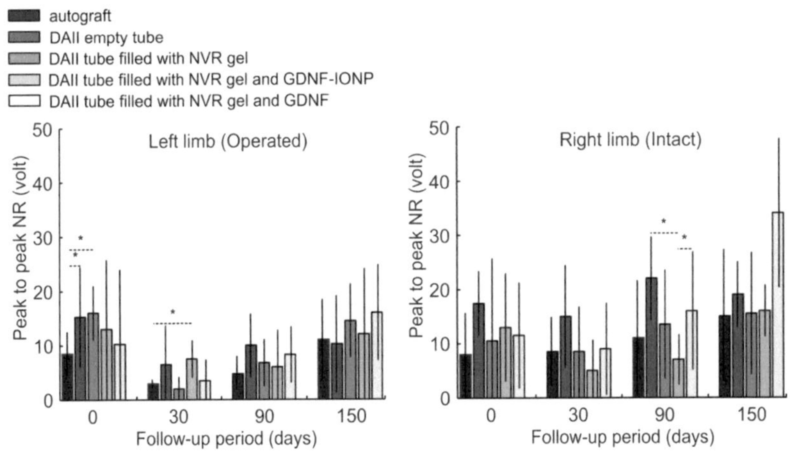

Figure 3. Comparison among the various treatments. Data were gathered from each limb separately (operated and intact) at four follow-up periods: 0 (pre-operatively), 30, 90, and 150 days. Values are presented as median ± mad statistical analysis: Mann-Whitney rank-sum test. *$p < 0.05$.

The same treatment seemed to provoke significant decrease in P2P amplitude at the intact limb but only during the first time point investigated. The difference was observed in relation to the empty tube group ($p < 0.05$), and the tube filled with NVR gel enriched with free GDNF ($p < 0.05$).

In order to normalize the results, we decided to subtract P2P measurement in the operated limb with the P2P measurement in the intact limb (**Figure 4**) in each of the four follow-up time points (0, 30, 90, and 150 days). The "0" value indicates similar amplitude in both intact and operated limbs. Since a dramatic decrease in amplitude is an indicator of neurological dysfunction, we assume that a large shift down from "0" is a marker for this pathology. As expected, at time point 0 (pre-operation), all groups exhibited an amplitude difference very close to "0." During the follow-up, most treatments exhibited a decrease at the amplitude. The most robust and sustained decrease was found at the empty tube group. Surprisingly, the rat group that was treated with tube filled with NVR gel enriched with GDNF-IONP had no decrease.

3.2. Morphological and morphometrical analysis

To determine how structural aspect of regenerated nerves is influenced by the choice of the dispositive used to repair the long gap, we performed morphological and morphometrical analysis on nerve regenerated inside the chitosan tube 5 months after the surgery. Beside only two animals showed regeneration in empty tube group, this group was excluded from morphometrical analysis. The morphometric quantification was carried out in the distal part of the regenerated nerve inside the tube (**Figure 5A**). Semi-thin section of regenerate nerves was used for the analysis and compared to the distal part of regenerated nerves of autograft group, our positive control (**Figure 5B–E**).

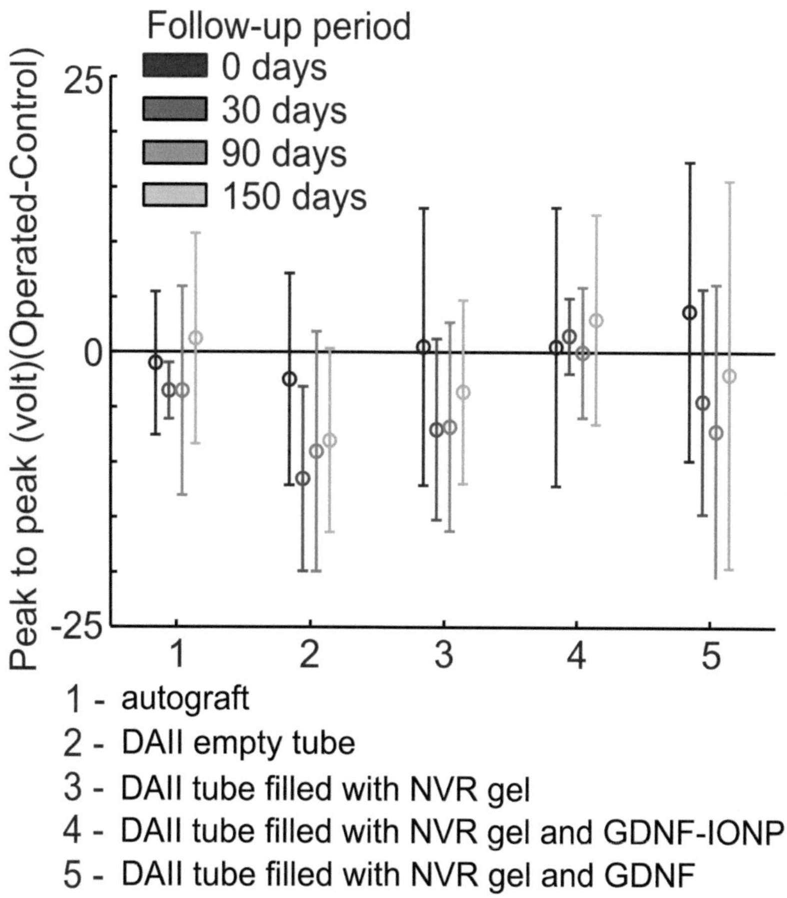

Figure 4. P2P measurement normalized to the intact limb. The graph shows the data obtained through the subtraction of P2P measurement in the operated limb with P2P measurement in the intact limb. The "0" value indicates similarity between the two limbs. Values are presented as median ± mad.

It is possible to notice the similarity among experimental groups in which nerve cross sections are organized in fascicles with small fibers.

The number of myelinated fibers is significantly higher in autograft group ($p \leq 0.001$) that remains the gold standard of regeneration (**Figure 6**); nevertheless, it is important to notice that in the experimental group repaired with the tube enriched with NVR gel and GDNF-IONP, the number of myelinated fibers is statistically significant higher ($p \leq 0.05$) respect to the group repaired with the conduit functionalized with NVR gel alone (**Figure 6**).

The myelin thickness parameter, recorded for the group repaired using a tube with NVR gel alone, shows a statistically relevant reduction compared to autograft group ($p \leq 0.05$). As regard the other parameters considered, it is interesting to note that there are no differences among the experimental groups.

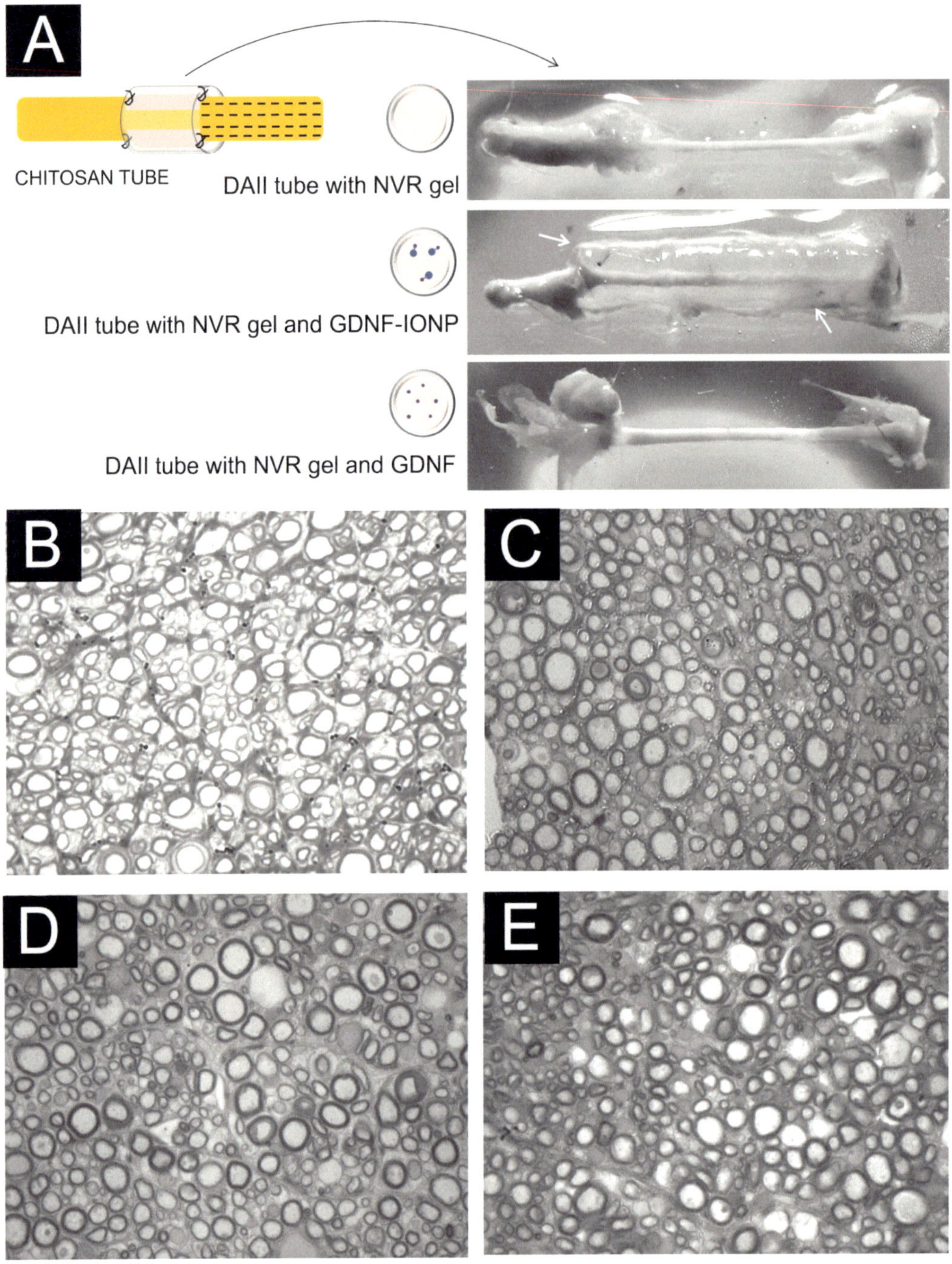

Figure 5. Morphological analysis of regenerated nerves inside chitosan tube. (A) Representative photos of sciatic rat regenerated nerves harvested inside chitosan tube 5 months after the implantation. (B–E) The figure shows representative semi-thin transversal nerve sections of regenerated sciatic nerves used for morphometrical analysis. For autograft group (B), the distal nerve is represented; for tube with NVR gel alone (C), tube with NVR gel enriched with GDNF-IONP (D) and tube with NVR gel enriched with free GDNF (E); the pictures refer to the nerve found inside the chitosan tube. For all the groups, a good regeneration is visible.

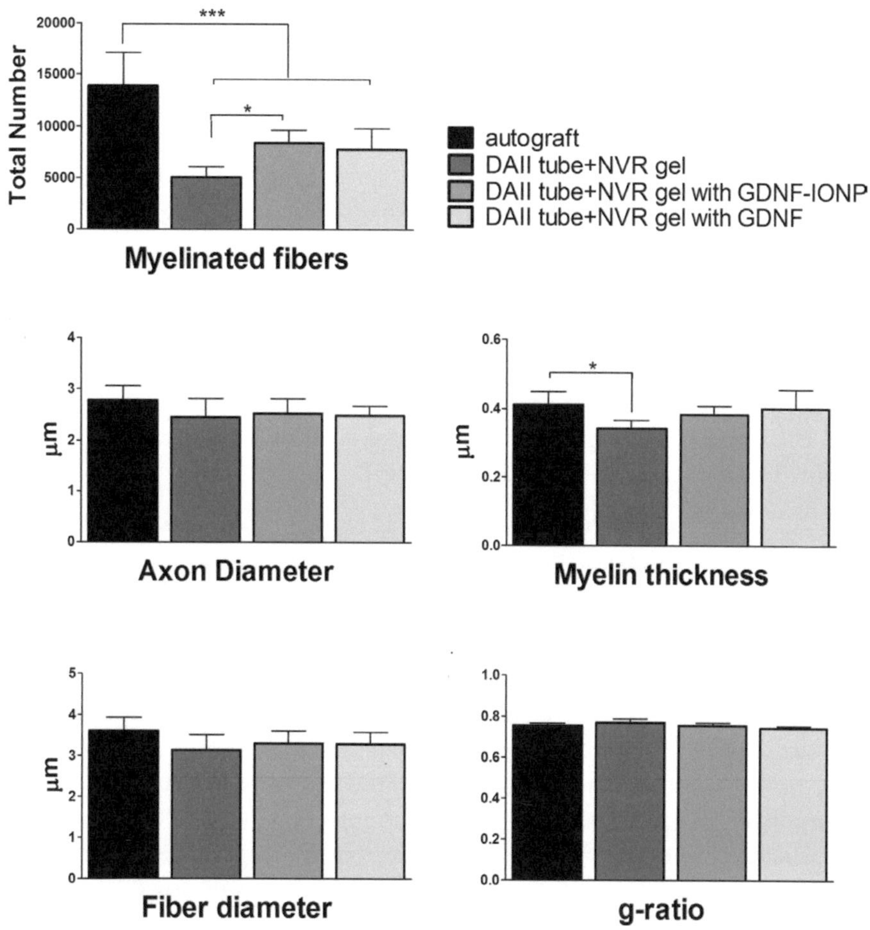

Figure 6. Morphometrical analysis of regenerated nerve inside the chitosan tube. The analysis was performed on semi-thin transverse sections of the regenerated nerves 5 months after implantation. For autograft group (used as control), the distal part of the nerve was analyzed; for all the other groups, the values in the graphs are referred to the distal regenerated nerve found inside the chitosan tube. The graphs show the total number of number of myelinated fibers, the axon diameter, the fiber diameter, the myelin thickness, and the g-ratio. In the filled tubes, there is less myelinated fibers respect to the autograft, but all the other parameters are comparable with the autograft. The two groups with GDNF factors, free or conjugated to iron-oxide nanoparticles, are similar. All data are presented as mean + SD. Statistical analysis: One-way ANOVA with post hoc Bonferroni Test. $^*p < 0.05$; $^{**}p < 0.01$; $^{***}p < 0.001$.

4. Discussion

A wide range of conduits with different internal design have been tested for reconstruction of peripheral nerves in animal models [28]. The efforts concern the attempt to replace the current gold standard, the autologous nerve graft, above all in the case of very serious injury and to avoid negative side-effects [2, 4]. Chitosan conduits, among the absorbable forms, have been shown to be able of supporting the peripheral nerve regeneration with values of recovery fully comparable to the surgical gold standard technique, with regard to sciatic nerve lesions of 10 mm in the rat animal model [7, 11]. Despite this, in case of nerve injuries with a considerable loss of substance, it is necessary to functionalize the conduit, to provide a filler, in order to prevent the collapse of the walls, and to provide a releasing system of neurotrophic molecules able to accelerate the regenerative processes [26]. In this study, we used the NVR gel, made of hyaluronic acid and laminin [13], as internal filler and for the release of the neurotrophic factor GDNF, which has been further enhanced by stabilizing its duration through a covalent conjugation with IONP [19].

We have previously demonstrated that the administration of this factor induces an early myelination in organotypic cultures of neonatal DRG [24]. The whole device has been proposed for the repair of a rat sciatic nerve lesion with severe loss of substance (15 mm). In this study, the nerve regeneration and the correct position of the device have been constantly monitored. Ultrasound observations and functional analysis have allowed us to carry out regular follow-up (30, 90, and 150 days after surgery).

The ultrasound analysis carried out on the implants revealed that none of these have undergone a shift from the correct position by the moment of the surgery. The two representative figures show the implanted tube at 20 and 90 days after surgery; at the experimental time 90 days, nerve regeneration appears through nerve fascicles.

Electrophysiological analysis on the five experimental groups demonstrates a nerve conductivity recovery (P2P amplitude) by the end of the follow-up period (150 days), when the tube enriched either with NVR gel or NVR gel and GDNF-IONP treatments displayed a complete recovery, as time 0, whether other treatments showed only a partial recovery. Normalizing the SSEP results of the operated limb (left) to the intact one (right), all groups showed a decrease after surgery followed by an increase, as expected, except for the group treated using a tube enriched with NVR gel and GDNF-IONP that did not showed any decrease during the follow-up period after surgery. These electrophysiological findings suggest that the treatment, in which a tube is enriched with NVR gel, is comparable to the autologous nerve graft treatment.

The morphological aspect of regenerated nerves when, after 5 months, rats were sacrificed and tubes explanted revealed for all the cases analyzed a good nerve regeneration.

The morphometric quantification was referred on the distal part of the regenerated nerve inside the tube. Semi-thin sections of nerves were used for the analysis and compared to the ones from the distal part of regenerated nerves of autograft group, our positive control. All the nerves revealed the same cross section structure: fascicles with small fibers, a typical nerve regeneration framework.

It was interesting to observe similar values for the measured parameters, such as axon diameter, myelin thickness, fiber diameter, and g-ratio, demonstrating an equal good regeneration, because compared with the gold standard, the autograft. The only parameter in which the superiority of our positive control was found was the total number of the fibers, although the group represented by the tube enriched with NVR gel and GDNF-IONP had a statistically significant greater value compared to the tubes enriched with only NVR gel.

5. Conclusions

In this work, we investigated the efficacy, in the repair of a nerve injury with a considerable loss of substance, of a conduit functionalized (i) with a factor that has been demonstrated to accelerate nerve regeneration, (ii) with a hydrogel, which has proven to be a good substrate for neurite outgrowth, (iii) and with the system of the IONP, able to stabilize the signaling of the factors. The analysis carried out has not shown a great improvement in nerve regeneration in animals treated with this functionalized device compared to the other devices investigated. IONPs are potentially a good candidate for nerve device enrichment; however, the best way to administer neurotrophic factors IONP in the tube needs further investigation, as well as the effects of their long time exposition.

Acknowledgements

This study was financially supported by the European Community's Seventh Framework Program (FP7-HEALTH-2011), grant agreement n° 278612 (BIOHYBRID). Altakitin SA (Lisbon, Portugal) supplied medical grade chitosan for nerve guides, which were produced by Medovent GmbH (Mainz, Germany).

Author details

Federica Fregnan[1*], Michela Morano[1], Ofra Ziv-Polat[2], Mira M. Mandelbaum-Livnat[3], Moshe Nissan[3], Tolmasov Michael[3], Akiva Koren[3], Tali Biran[3], Yifat Bitan[3], Evgeniy Reider[3], Mara Almog[3], Nicoletta Viano[1], Shimon Rochkind[3], Stefano Geuna[1] and Abraham Shahar[2]

*Address all correspondence to: federica.fregnan@unito.it

1 Department of Clinical and Biological Sciences, Cavalieri Ottolenghi Neuroscience Institute, University of Turin, Turin, Italy

2 NVR Research Ltd., Ness-Ziona, Israel

3 Division of Peripheral Nerve Reconstruction, Department of Neurosurgery, Research Center for Nerve Reconstruction, Tel Aviv Sourasky Medical Center, Tel Aviv University, Tel-Aviv, Israel

References

[1] Ciaramitaro P, et al. Traumatic peripheral nerve injuries: Epidemiological findings, neuropathic pain and quality of life in 158 patients. Journal of the Peripheral Nervous System. 2010;**15**(2):120–127

[2] Battiston B, et al. Nerve repair by means of tubulization: Literature review and personal clinical experience comparing biological and synthetic conduits for sensory nerve repair. Microsurgery. 2005;**25**(4):258–267

[3] Gu XS, et al. Construction of tissue engineered nerve grafts and their application in peripheral nerve regeneration. Progress in Neurobiology. 2011;**93**(2):204–230

[4] Konofaos P, Halen JPV. Nerve repair by means of tubulization: Past, present, future. Journal of Reconstructive Microsurgery. 2013;**29**(3):149–163

[5] Isaacs J, Browne T. Overcoming short gaps in peripheral nerve repair: Conduits and human acellular nerve allograft. Hand (N Y). 2014;**9**(2):131–137

[6] Gnavi S, et al. The use of chitosan-based scaffolds to enhance regeneration in the nervous system. Tissue Engineering of the Peripheral Nerve: Biomaterials and Physical Therapy. 2013;**109**:1–62

[7] Shapira Y, et al. Comparison of results between chitosan hollow tube and autologous nerve graft in reconstruction of peripheral nerve defect: An experimental study. Microsurgery. 2016;**36(8)**:664–671

[8] Simoes MJ, et al. In vitro and in vivo chitosan membranes testing for peripheral nerve reconstruction. Acta Medica Portuguesa. 2011;**24**(1):43–52

[9] Yuan Y, et al. The interaction of Schwann cells with chitosan membranes and fibers in vitro. Biomaterials. 2004;**25**(18):4273–4278

[10] Freier T, et al. Controlling cell adhesion and degradation of chitosan films by N-acetylation. Biomaterials. 2005;**26**(29):5872–5878

[11] Haastert-Talini K, et al. Chitosan tubes of varying degrees of acetylation for bridging peripheral nerve defects. Biomaterials. 2013;**34**(38):9886–9904

[12] Stenberg L, et al. Nerve regeneration in chitosan conduits and in autologous nerve grafts in healthy and in type2 diabetic Goto-Kakizaki rats. European Journal of Neuroscience. 2016;**43**(3):463–473

[13] Shahar A, Nevo Z, Rochkind S. Cross-linked hyaluronic acid-laminin gels and use thereof in cell culture and medical implants. U.S. patent. Editor. 2001: US

[14] Boyd JG, Gordon T. Neurotrophic factors and their receptors in axonal regeneration and functional recovery after peripheral nerve injury. Molecular Neurobiology. 2003;**27**(3):277–323

[15] Hoke A, et al. Glial cell line-derived neurotrophic factor alters axon Schwann cell units and promotes myelination in unmyelinated nerve fibers. Journal of Neuroscience. 2003;**23**(2):561–567

[16] Santos D, et al. Dose-dependent differential effect of neurotrophic factors on in vitro and in vivo regeneration of motor and sensory neurons. Neural Plasticity. 2016;**2016**:4969523

[17] Gordon T. The role of neurotrophic factors in nerve regeneration. Neurosurgical Focus. 2009;**26**(2):E3

[18] Hoke A, et al. A decline in glial cell-line-derived neurotrophic factor expression is associated with impaired regeneration after long-term Schwann cell denervation. Experimental Neurology. 2002;**173**(1):77–85

[19] Morano M, et al. Nanotechnology versus stem cell engineering: In vitro comparison of neurite inductive potentials. International Journal of Nanomedicine. 2014;**9**:5289–5306

[20] Chertok B, et al. Iron oxide nanoparticles as a drug delivery vehicle for MRI monitored magnetic targeting of brain tumors. Biomaterials. 2008;**29**(4):487–496

[21] Corot C, et al. Recent advances in iron oxide nanocrystal technology for medical imaging. Advanced Drug Delivery Reviews. 2006;**58**(14):1471–1504

[22] Maier-Hauff K, et al. Intracranial thermotherapy using magnetic nanoparticles combined with external beam radiotherapy: Results of a feasibility study on patients with glioblastoma multiforme. Journal of Neuro-Oncology. 2007;**81**(1):53–60. Epub 2006 Jun 14

[23] Ziv-Polat O, Margel S, Shahar A. Application of iron oxide anoparticles in neuronal tissue engineering. Neural Regeneration Research. 2015;**10**(2):189–191. DOI: 10.4103/1673-5374.152364

[24] Ziv-Polat O, et al. The role of neurotrophic factors conjugated to iron oxide nanoparticles in peripheral nerve regeneration: In vitro studies. BioMed Research International. 2014;**2014**:267808

[25] Geuna, S., et al., Appreciating the difference between design-based and model-based sampling strategies in quantitative morphology of the nervous system. J Comp Neurol. 2000;**427**(3):333–9

[26] Meyer C, et al. Chitosan-film enhanced chitosan nerve guides for long-distance regeneration of peripheral nerves. Biomaterials. 2016;**76**:33–51

[27] Meyer C, et al. Peripheral nerve regeneration through hydrogel-enriched chitosan conduits containing engineered schwann cells for drug delivery. Cell Transplantation. 2016;**25**(1):159–182

[28] Tos P, et al. Future perspectives in nerve repair and regeneration. International Review of Neurobiology. 2013;**109**:165–192